SpringerBriefs in Public Health

For further volumes:
http://www.springer.com/series/10138

Veda Eswarappa · Sujata K. Bhatia

Naturally Based Biomaterials and Therapeutics

The Case of India

 Springer

Veda Eswarappa
School of Engineering
   and Applied Sciences
Harvard University
Cambridge, MA
USA

Sujata K. Bhatia
School of Engineering
   and Applied Sciences
Harvard University
Cambridge, MA
USA

ISSN 2192-3698   ISSN 2192-3701   (electronic)
ISBN 978-1-4614-5385-7   ISBN 978-1-4614-5386-4   (eBook)
DOI 10.1007/978-1-4614-5386-4
Springer New York Heidelberg Dordrecht London

Library of Congress Control Number: 2012945478

© The Author(s) 2013
This work is subject to copyright. All rights are reserved by the Publisher, whether the whole or part of the material is concerned, specifically the rights of translation, reprinting, reuse of illustrations, recitation, broadcasting, reproduction on microfilms or in any other physical way, and transmission or information storage and retrieval, electronic adaptation, computer software, or by similar or dissimilar methodology now known or hereafter developed. Exempted from this legal reservation are brief excerpts in connection with reviews or scholarly analysis or material supplied specifically for the purpose of being entered and executed on a computer system, for exclusive use by the purchaser of the work. Duplication of this publication or parts thereof is permitted only under the provisions of the Copyright Law of the Publisher's location, in its current version, and permission for use must always be obtained from Springer. Permissions for use may be obtained through RightsLink at the Copyright Clearance Center. Violations are liable to prosecution under the respective Copyright Law.
The use of general descriptive names, registered names, trademarks, service marks, etc. in this publication does not imply, even in the absence of a specific statement, that such names are exempt from the relevant protective laws and regulations and therefore free for general use.
While the advice and information in this book are believed to be true and accurate at the date of publication, neither the authors nor the editors nor the publisher can accept any legal responsibility for any errors or omissions that may be made. The publisher makes no warranty, express or implied, with respect to the material contained herein.

Printed on acid-free paper

Springer is part of Springer Science+Business Media (www.springer.com)

Preface

Today, more than ever before, there is a pressing need for biomedical innovations to address the plethora of health problems afflicting the world. Particularly in developing countries like India—where communicable diseases, non-communicable diseases, and injuries all contribute substantially to overall morbidity and mortality—there is no panacea that will easily pave the way to improved health outcomes. Different cultural, economic, infrastructural, and other factors also prevent many interventions currently popular in the developed world from being sufficiently effective in the developing world. For example, despite having far more people, nominal GDP in India and China in 2009 was less than one-fifth that of the United States and Western Europe,[1] a statistic that makes it clear that many high-tech, expensive procedures conducted in the US might not be as fitting for use in India.

The lack of suitable options currently available to improve health outcomes in developing countries like India presents a tremendous opportunity for less-traditional approaches, including those that utilize naturally based biomaterials and therapeutics, an area that has traditionally been overlooked but has also demonstrated impressive potential for health applications in recent years. This book seeks to explore precisely these kinds of applications, which can enable countries like India to access more effective, inexpensive treatments while also taking more ownership of their healthcare technologies and innovations.

The book is divided into four distinct chapters. Chapter 1 provides some background and context for the rest of the book, giving a brief introduction to global health, biomaterials, and therapeutics. It starts by surveying the current state of health worldwide before focusing specifically on India and the challenges this populous, diverse country faces. Then it covers the significance of biomaterials and therapeutics and their promise in the quest to improve world health.

Chapter 2 introduces nine natural resources with potential medical applications. These resources—bamboo, banana, coconut, jackfruit, jute, rice, silk, soy, and

---

[1] Hamilton, D. and J. Quinlan. "The Transatlantic Economy 2011: Annual Survey of Jobs, Trade and Investment between the United States and Europe." *Center for Transatlantic Relations*, 2011.

tamarind—are commonly found in India as well as in certain other regions of the world. This chapter explores their various properties and current applications.

Chapter 3 delves into some of the most important potential biomaterial and therapeutic applications for each of these nine crops. Such applications vary from drug-eluting stents to treatments that eliminate intestinal parasites. Some of these applications are already in use today or have been used in times previously, such as coconut water as an alternative to popular contemporary intravenous fluid treatments. Other applications—such as the use of cross-linked carboxymethyl jackfruit starch as a tablet disintegrant—require additional exploration and research before they can be proven effective and safely used for humans. Each of these biomedical applications has demonstrated abilities to combat health concerns that are highly relevant to India, as well as other regions of the world.

Chapter 4, the final chapter of this book, summarizes the primary themes of the previous chapters and discusses the implications of these key points. It then raises and addresses several of the most compelling challenges and objections to using naturally sourced biomaterials and therapeutics for health applications, particularly in developing countries. Finally, it presents several recommendations and lays out a basic plan to move forward and harness the potential of naturally derived biomaterials and therapeutics to maximize their benefit in health applications.

Altogether, this book aims to help readers better understand some of the key health concerns facing countries like India and how naturally derived biomaterials and therapeutics could help substantially alleviate many of these problems. This is an exciting time for such biomaterial and therapeutic research, and hopefully further exploration of bio-derived materials will yield even greater benefits for world health outcomes in years to come.

Contents

Part I

1   Introduction to Global Health, Biomaterials, and Therapeutics . . .   3
The State of Health Worldwide.   3
The State of Health in India   6
Biomaterials and Therapeutics: A Role in World Health   10
References   12

2   Natural Resources with Potential for Health Applications   15
Bamboo   15
Banana.   16
Coconut   17
Jackfruit.   18
Jute   19
Rice.   20
Silk   21
Soy   21
Tamarind   22
References   23

Part II

3   Biomaterial and Therapeutic Applications.   27
Bamboo Applications.   27
      Blood Purification   27
      Bone Repair or Replacement.   28
      Neuroprotector   30
Banana Applications   31
      Burn Dressing.   31
      Inhibition of Viral Transmission   32
      Diabetes Treatment   34

Bone Grafting...................................... 35
Anti-Ulcerogenic................................... 36
Iron Absorption Promoter............................ 37
Coconut Applications..................................... 37
Obesity and Diabetes Treatment 37
Helminth Treatment................................. 39
Ulcer Treatment 40
Intravenous Treatment 40
Jackfruit Applications 41
Protease Inhibitor 41
Tablet Disintegrant 43
Leukemia Treatment 43
Jute Applications.. 45
Wound Dressing 45
Arsenic Protection.................................. 45
Digesta Viscosity Elevation 47
Rice Applications 48
Protein Production.................................. 48
Multifunctional Excipient............................ 50
Porous Scaffolds 51
Sutures... 52
Silk Applications.. 54
Corneal Grafts..................................... 54
Optical Devices.................................... 55
Conduits for Nerve Repair 57
Bone Regeneration.................................. 58
Drug Delivery..................................... 59
Cartilage and Ligament Repair 61
Multi-Functional Stents 62
Soy Applications.. 63
Bone Regeneration.................................. 63
Inflammation Treatment.............................. 65
Drug Delivery..................................... 66
Tamarind Applications................................... 67
Drug Delivery..................................... 67
Opthalmic Applications 69
Anti-Malarial 71
Anti-Obesity...................................... 72
Antihelminthic 72
References 73

Part III

4 Implications, Challenges, and Recommendations   83
   Summary   83
   Objections and Challenges   84
   Moving Forward   87
   References   89

Part I

Chapter 1
Introduction to Global Health, Biomaterials, and Therapeutics

The State of Health Worldwide

The general health status of the world has improved substantially over the past two centuries, likely due to a combination of better nutrition, economic growth, public health developments, and improved medical treatments. The last half century alone has seen tremendous improvements, with life expectancies in India and China soaring by approximately 30 years since 1950.[1] Nevertheless, the world is still far from achieving the Millennium Development Goals, eight largely health-related international development objectives established in 2000. As scientists and engineers, it is critical that we assess the state of these and other health goals to determine where the needs are most pressing and where we should direct our research efforts to have a meaningful impact.

Childhood and maternal morbidity and mortality are still major problems today. With approximately 115 million children younger than 5 years underweight and about 178 million children physically stunted by WHO standards, undernutrition is still a significant problem, particularly in Africa and Asia. Child mortality also remains a problem. Despite the fact that global child mortality rates have fallen by about one-third since 1990, mortality levels in Africa and low-income countries are still too high, given that they are above the average 1990 levels. Diarrheal diseases and pneumonia are the two leading causes of death in children under 5 years of age, responsible for about one-third of those deaths. Although maternal mortality rates have fallen over time, 99 % of maternal deaths in 2008 occurred in developing countries and maternal mortality rates in Africa remained disturbingly high.

---

[1] Cutler et al. 2006.

V. Eswarappa and S. K. Bhatia, *Naturally Based Biomaterials and Therapeutics*, SpringerBriefs in Public Health, DOI: 10.1007/978-1-4614-5386-4_1, © The Authors(s) 2013

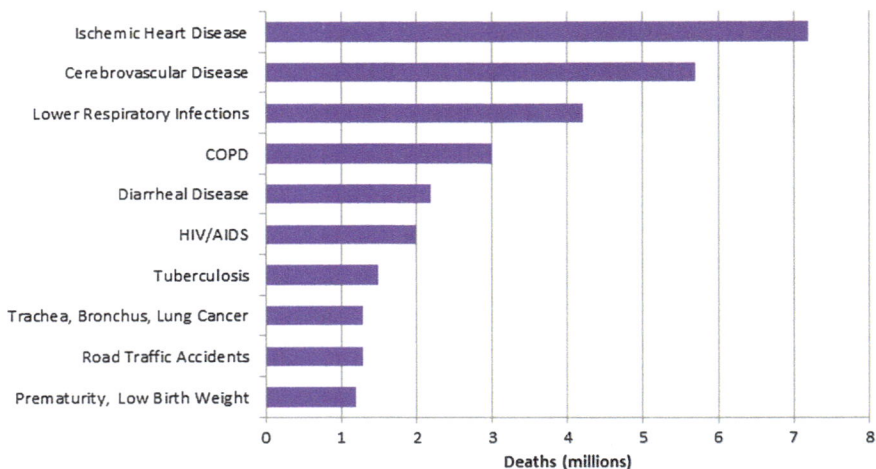

Fig. 1.1 Top ten causes of death worldwide in 2004 (Data from WHO, http://www.who.int/ healthinfo/global_burden_disease/GBD_report_2004update_full.pdf)

In addition, child health interventions—such as oral rehydration therapy for diarrhea or antibiotics for pneumonia—are often not widely available and particular immunization efforts (especially in Africa) have been less successful than desired.[2]

Infectious diseases may often be overlooked in the United States, but as shown in Fig. 1.1, they still have a substantial impact elsewhere in the world. Although over 780,000 people worldwide died from malaria in 2009, many countries experienced decreasing rates of the illness, with 32 countries reporting a drop in confirmed malaria cases of more than 50 % from 2000 to 2009. New tuberculosis incidences have been rising somewhat, though such growth is largely due to population growth, and mortality due to tuberculosis has decreased by over a third since 1990. However, tuberculosis rates in Africa are declining especially slowly, and multidrug-resistant strains of tuberculosis will likely pose a great problem for the world in the future. New HIV infections in 2009 were 19 % lower than a decade prior, but there are still over 33 million people infected and treatment-coverage is quite low in low- and middle-income countries, at just 36 %.[3] Over 1 billion people have a neglected tropical disease and about 2 billion others are at risk, with these diseases heavily concentrated in poor areas of tropical and subtropical regions.[4] While cases of leprosy and dracunculiasis have plummeted by over 90 % since 1989, dengue outbreaks have been increasing in number and expanding geographically.[5]

---

[2] World Health Organization 2011a.
[3] Ibid.
[4] Kaiser Family Foundation 2011.
[5] World Health Organization 2011a.

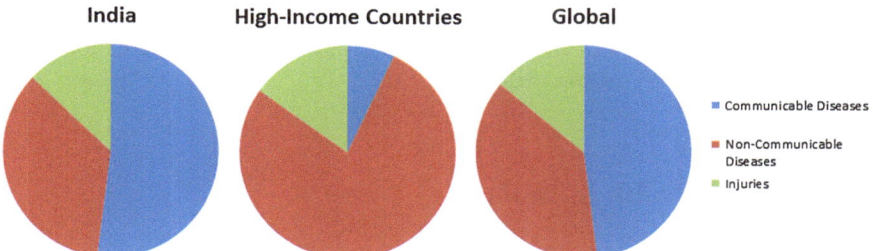

Fig. 1.2 Comparison of distribution years of life lost due to communicable diseases, non-communicable diseases, and injuries in 2008 (From WHO, http://www.who.int/whosis/whostat/EN_WHS2011_Full.pdf)

Non-communicable diseases have been having an increasingly large effect on health worldwide over the past few decades (Fig. 1.2). Approximately 36 million deaths in 2008 were due to non-communicable diseases including cancers, cardiovascular diseases, chronic respiratory diseases, and diabetes.[6] Cardiovascular disease is the leading cause of death worldwide, causing over three times the total number of annual deaths from HIV/AIDS, malaria, and tuberculosis combined. Despite its frequent characterization as a "problem of wealthy, industrialized nations", cardiovascular disease has a substantial impact on both developed and developing countries. It is now responsible for nearly 3 of every 10 deaths in low- and middle-income countries and over 80 % of all cardiovascular disease-related deaths now take place in such nations.[7]

Injuries—both intentional and unintentional—are responsible for about 9 % of deaths worldwide. Over 5 million people each year die directly as a result of their injuries, while millions of others suffer from injuries. The global burden of injury falls heavily upon the world's poor, as they are more likely to live, travel, and work in unsafe environments but less likely to benefit from prevention efforts or have sufficient access to good treatment or rehabilitation facilities.[8]

Infrastructural factors also play an important role in health worldwide. Global access to improved drinking-water sources improved from 77 to 87 % between 1990 and 2008, but over 880 million people (most of whom were living in rural areas) were still dependent on unimproved water sources in 2008. Sanitation also poses a major problem to global health, since approximately 2.6 billion people did not have access to improved sanitation facilities in 2008. More than 1.1 billion of those people lived without access to any toilets or sanitation facilities. In addition, relatively high costs and low availability have made essential medicines a problem in much of the world. Since only 42 % of public sector health facilities in low- and middle-income countries had particular generic medicines (in comparison to 64 %

---

[6] Ibid.

[7] Fuster and Kelly 2010.

[8] World Health Organization 2011a.

of private facilities), many patients have to buy pharmaceuticals from the private sector, where they pay an average of 6.3 times the cost of the international reference price for those generic medicines.[9]

The State of Health in India

Many people in the western world associate Indians with high quality medical care. This seems like a reasonable assumption given the number of doctors India produces each year (there were over 30,000 medical college graduates in 2010 alone),[10] many of whom immigrate to western countries. In the United States, 1 in 20 doctors is of Indian origin.[11] That figure jumps to more than one in ten for Canadian doctors, and an astounding one in five for Australian doctors.[12]

India's growing reputation as a destination for high-quality medical tourism has only bolstered the association between Indians and quality healthcare. By 2013, India's medical tourism industry could be generating revenues of approximately US$3 billion.[13] The Narayana Hrudalaya Heart Hospital in Bangalore, one of India's largest cities, is one example of a premier health facility that is able to attract the attentions of many foreigners. Each year, this hospital performs more than 3.5 times as many CABG (Coronary Artery Bypass Graft) surgeries as Boston's Mass General Hospital. In addition, it boasted twice as many cardiac surgeons as Minneapolis's Mayo Clinic, with each surgeon averaging nearly twice as many surgeries as their counterparts at the Cleveland Clinic in Ohio.[14]

However, the overall health status in India itself paints a less rosy picture. India's population is very diverse and densely packed—with over 1 billion people occupying less than 3 % of the world's area[15]—two factors which make it challenging to devise a simple way to categorically improve the nation's health. The South Asian country claims a significant fraction of the world's burden of disease, with 18 % of global deaths and 20 % of disability-adjusted life-years as of 2009.[16] Recent figures also indicate that 1 in 5 maternal deaths and 1 in 4 child deaths worldwide happen in India.[17, 18]

[9] Ibid.

[10] Duttagupta 2011.

[11] Knox 2007.

[12] Duttagupta 2011.

[13] India Knowledge @ Wharton 2011.

[14] Khanna et al. 2005.

[15] World Health Organization 2006.

[16] World Health Organization 2009.

[17] UNICEF 2009.

[18] UNICEF 2008.

As of 2009, the life expectancy at birth in India had increased to 65 years from 57 years in 1990 and 29 years in 1930.[19, 20] Although this is a substantial improvement, it still lags behind the 2009 global total life expectancy of 68 and is nowhere near the 80-year life expectancy of high income countries. These trends from 1990 to 2009 are shown in Fig. 1.3. Similar trends can be observed in the country's mortality rates, despite the country's tremendous progress in the area since achieving independence in 1947. As shown in Fig. 1.4, India's infant mortality rate declined significantly—from 84 deaths per 1000 live births in 1990 to 50 in 2009—but remained above the global total of 42 and the high-income rate of 6. India's adult mortality rate dropped as well, from 274 deaths per 1000 people aged 15–60 to 212 deaths, but stayed much higher than the global average of 176 and the high-income value of 88.[21]

The distribution of years of life lost by broader causes in India follows a pattern demonstrated by most lower-income countries, where communicable diseases have the largest share. As shown in Fig. 1.2, in 2008, communicable diseases accounted for 52 % of such years in India, while non-communicable diseases were responsible for 35 % and injuries represented the remaining 13 %. This is not remarkably different from the global total of 48 % communicable, 38 % non-communicable, and 14 % injuries, but it is striking in contrast to the high-income countries. There, communicable diseases account for the smallest share at just 7 %, while non-communicable diseases account for 77 % and injuries represent 15 %.[22]

Trauma injuries are a sizeable cause of mortality and morbidity in India. As of 2008, about one-fifth of all emergency-related visits in India were related to trauma. By 2020, road accidents will account for over half a million deaths in the country, according to the World Health Organization.[23]

Compared to the high-income country average of 28.6 (physicians per 10,000 people), or the global total of 14, India's physician density is quite low. At just six physicians per 10,000 in the population, it falls well below the global median of 11.5 as well.[24] Such a low density reflects India's staggering physician shortage of over 50 %. Other indicators of medical facilities in India do not paint a much better picture, with the density of hospital beds in the population just 40 % of the density in China but the average number of doctors per bed still under three-quarters of the level in the United States.[25]

The dearth of physicians and medical physicians in one of the world's most populous countries is further accentuated by its relatively low health expenditures.

---

[19] World Health Organization 2011b.

[20] Fogel 2004.

[21] World Health Organization 2011b.

[22] World Health Organization 2011c.

[23] Das et al. 2008.

[24] World Health Organization 2010a.

[25] Duttagupta 2011.

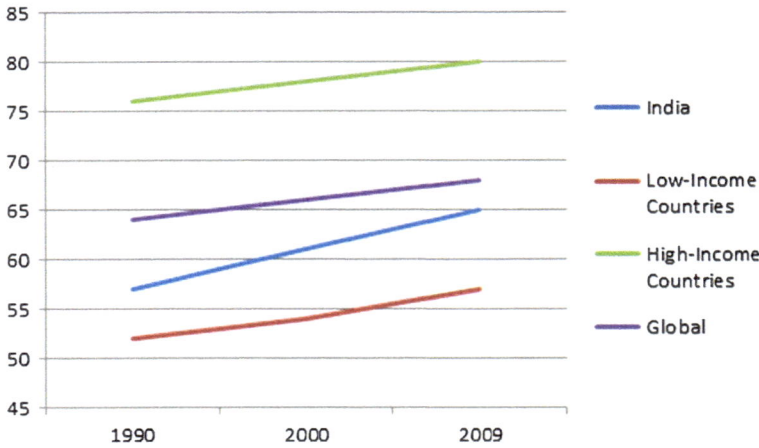

Fig. 1.3 Comparison of trends in life expectancy, sampled in 1990, 2000, and 2009 (Data from WHO, http://www.who.int/whosis/whostat/EN_WHS2011_Full.pdf)

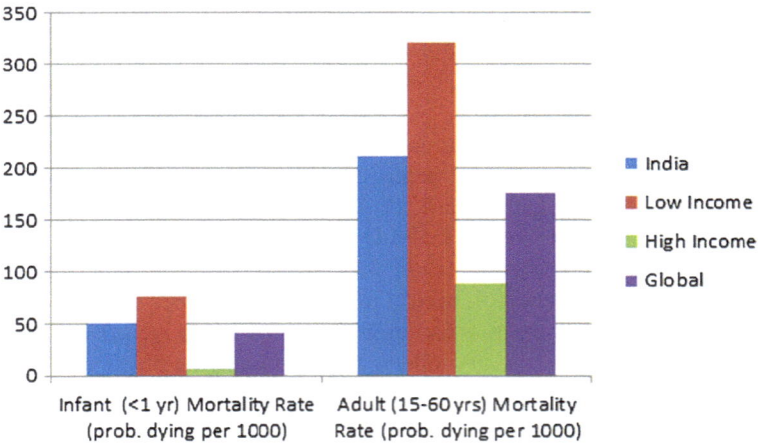

Fig. 1.4 Comparison of infant and adult mortality rates in 2009 (Data from WHO, http://www.who.int/whosis/whostat/EN_WHS2011_Full.pdf)

India's total expenditure on health equaled 4.2 % of GDP, actually declining from the 2000 level of 4.6 %. This level was less than half of the global total expenditure on health as a percentage of GDP and a measly 38 % of the GDP fraction high income countries spent. It is also lower than the 5.4 % of GDP that low-income countries overall devoted to health expenditures.[26]

---

[26] World Health Organization 2010b.

Time Period	Materials	Applications	Investigators/Introducers
~2000 BC	Elephant tusks, walrus teeth, linen	Artificial legs, teeth, ears, and sutures	Egyptians
~1800 BC	Wood	Artificial legs, teeth, ears, and sutures	Indians, Chinese
~200 BC	Gold	Wires for fractures	Greeks
1759 AD	Wooden peg and thread	Brachial artery	Hallowel
1860 AD	Catgut	Suture	J. Lister
1893 AD	Carbon steel	Bone plates and screws	W.A. Lane
1952 AD	Cloth	1st blood vessel replacement	Vorhees et al.
1997 AD	TransCyte®	1st FDA-approved synthetic skin	Advanced Tissue Sciences, USA
2002 AD	InFuse™	1st FDA-approved bone growth factor	Medtronic Sofamor Danek, USA
2006 AD	AbioCor®	1st FDA-approved totally implanted artificial heart	Abiomed, Inc., USA

Fig. 1.5 Sample biomaterial uses over time (Data from Ramakrishna et al. 2010)

This low level of health expenditures is compounded by atypically high out-of-pocket spending. Over three-quarters of health spending in India is private, the majority of which is paid out-of-pocket by individual households.[27] Such a high financial burden exacerbates inequalities in healthcare access and seriously affects tens of millions of Indians, driving about 2.2 % of India's population into poverty each year.[28]

Nevertheless, it is important to note that these figures do not explain or even note the tremendous variations in health within India. Caste, education, gender, geography, and wealth all play important roles in determining health in India.[29] In 2005–2006, the poorest quintile had an infant mortality rate that was 2.4 times that of the richest quintile.[30] Interstate life expectancy differences in India are far more pronounced than in the US or in China (with provinces).[31, 32] For example, in the southern state of Kerala, life expectancy is 74 years, while it is just 56 years in Madhya Pradesh.[33] Regardless, the state of health in India today leaves much to be desired. Hence, there is considerable interest in exploring different methods by which to improve health outcomes in the country.

---

[27] Balarajan et al. 2011.

[28] World Health Organization 2006.

[29] Subramanian et al. 2008.

[30] International Institute for Population Sciences and Macro International 2007.

[31] Burd-Sharps et al. 2008.

[32] World Health Organization 2008.

[33] Balarajan et al. 2011.

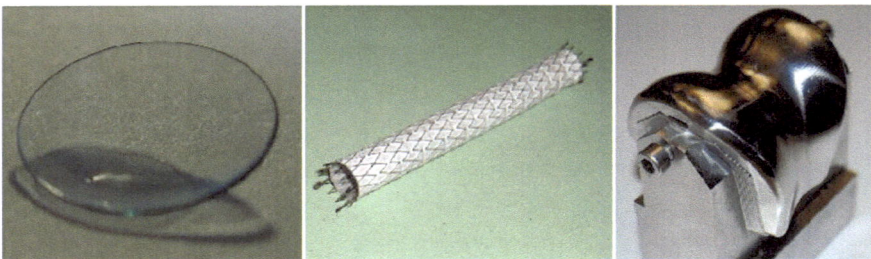

Fig. 1.6 Examples of biomaterial devices made of different materials from left: a contact lens, a stent device, and a knee prosthesis (From Wikimedia Commons Users Bryan Tong Minh, Haudebourg, and Sescoi CAD/CAM, respectively)

Biomaterials and Therapeutics: A Role in World Health

One widely accepted definition for a biomaterial is "a nonviable material used in a medical device, intended to interact with biological systems". Successful biomaterials must also be biocompatible, meaning that they must be able to "perform with an appropriate host response in a specific application".[34] Examples of biomaterial applications today include heart valve prostheses, contact lenses, and scaffolding for bone growth. Although many biomaterial applications can sound quite fancy or complex, biomaterials actually have a long history—ancient Egyptians used linen for sutures[35] while ancient Romans, Chinese, and Aztecs used biomaterials like gold as dental implants.[36] More recently, however, we have focused on engineering more complex biomaterials to best suit the needs of patients. A selection of biomaterial uses from ancient to current times is presented in Fig. 1.5.

Therapeutics aim to serve and care for a patient by preventing a specific condition or treating and managing a particular problem.[37] Examples of therapeutics include the delivery of various hormones, medicines, or vaccines such as a hydrogel to deliver a chemotherapy drug or a microbicide containing antiviral agents.

As shown in Fig. 1.6, biomaterials and therapeutics can be made up of a variety of materials, including ceramics, metals, or petroleum derivatives. They can also be based on natural materials such as bamboo or silk. In our increasingly modernizing world, it may seem strange to use natural materials in lieu of our latest, flashiest technologies.

Why put bamboo or silk in your body when you could have silicone or titanium? There are actually a number of situations where naturally based

---

[34] Ratner et al. 2004.

[35] Buntz 2011.

[36] Anderson et al. 2004.

[37] Encyclopaedia Britannica 2012.

biomaterials and therapeutics have demonstrated the capacity to perform just as well as—if not better than—other, seemingly state-of-the-art materials.

Although there is some discussion about therapeutics for the developing world today, discussions about biomaterials tend to revolve around applications for the developed world.[38] One estimate valued the global market for biomaterials at $25.6 billion in 2008 and expected the market to surge to $64.7 billion in 2015. The same source expected the biomaterial device market size to grow from $115.4 to $252.7 billion over the same time period.[39] However, such estimates are largely based on projections and needs for industrialized countries. Despite the fact that 80 % of the world's population resides in the developing world, the needs and applications of biomaterials for such populations are looked at with far less frequency and are rarely considered with the same sense of importance.[40]

There are a number of advantages that biomaterials and therapeutics possess with respect to applications in the developing world. Cost is a critical consideration, particularly given the fact that many biomaterial implants in the United States cost thousands of dollars and many therapeutic costs are very high as well, while total health expenditure in low-income countries averaged just $22 per capita in 2006. If biomaterial and therapeutic costs were kept sufficiently low, not only would more people in the developing world be able to access necessary treatments, but it is less likely that improper device reuse or repair would occur as well. On a related note, it is also important to develop biomaterial devices that can be safely cleaned and effectively reused. Biomaterials should also be mechanically robust, particularly since repair or replacement can be prohibitively expensive or infeasible due to lack of trained staff or facilities in some developing areas.

In addition, biomaterials and therapeutics for medical applications should be sufficiently easy to use. Medical personnel in some developing regions might not have access to state-of-the-art facilities or training, so needlessly complicated biomaterials would serve little purpose. Since many developing countries are situated in warmer climates and might not be able to properly maintain the cold chain at all times, developing the biomaterials and therapeutics to be stable at high temperatures is also important. These biomaterials and therapeutics should also pose minimal biohazard risk upon disposal because many medical facilities in the developing world might not be properly equipped for safe biohazard disposal.[41] Biomaterials for developing world applications should also be designed with a lower susceptibility to infection, since they may be used in environments where contamination rates are higher due to challenges in sterility maintenance.[42]

Recent years have seen increased recognition of the promise that naturally based biomaterials and therapeutics possess, particularly for use in developing

---

[38] Moussy 2010.

[39] Markets and Markets 2011.

[40] Moussy 2010.

[41] Ibid.

[42] Rosenthal et al. 2006.

countries such as India. Although most synthetic or high-tech solutions may sound good on paper, many have proven to be less than perfect, encountering issues with immune responses, mechanical stresses, or other factors.

Naturally based materials have the potential to help address some of these problems. Oftentimes, natural materials are simpler to work with and offer unparalleled biocompatibility and biomimicry, a state that fully or partially "mimics or inspires the biological mechanism".[43] If the biomaterial or therapeutic is able to compatibly interact with the body and avoid any signs of distress or immune response, it is likely that the patient will encounter fewer complications. This is of particular importance to developing countries, where facilities to deal with highly complex technologies might be less common and environmental barriers to achieve a biocompatible state might be higher than in developed countries. Such obstacles are exacerbated by the shortage of trained staff and adequately equipped medical facilities in India described above.

Many naturally based biomaterials and therapeutics can be easily manipulated with respect to chemical, mechanical, and physical properties. This is critical for honing drugs and devices to specific applications in the body and ensuring optimal results. This characteristic flexibility also allows bio-derived materials to be adjusted to a variety of targets and can enable one general material to successfully adapt to different environments. In addition, most bio-derived materials are renewably sourced, so there is far less environmental concern than with other biomaterial and therapeutic devices, particularly those that are sourced from petroleum.

To date, some studies have demonstrated that naturally derived materials have promise for the biomedical field. Therefore, the next chapter will discuss crops with potential for biomaterial or therapeutic applications in the developing world, with specific reference to India.

References

Anderson D et al (2004) Smart biomaterials. Science 305:1923–1924
Balarajan Y et al (2011) Health care and equity in India. The Lancet 377:12
Buntz B (2011) The legacy of biomaterials in medicine. Med Device Diagn Ind. http://www.mddionline.com/article/legacy-biomaterials-medicine
Burd-Sharps S et al (2008) The measure of America: American human development report 2008–09. Columbia University Press, Social Science Research Council, New York
Cutler D et al (2006) The determinants of mortality. J Econ Perspect 20:97–120
Das A et al (2008) White paper on academic emergency medicine in India: INDO-US joint working group (JWG). J Assoc Physicians India 56:789–797
Duttagupta I (2011) Indian healthcare: stop the brain drain of doctors. The Economic Times. http://articles.economictimes.indiatimes.com/2011-08-20/news/29909305_1_indian-doctors-physicians-of-indian-origin-british-association

---

[43] Ramakrishna et al. 2010.

Encyclopaedia Britannica (2012) Therapeutics. Encyclopaedia Britannica Online Academic Edition. http://www.brittanica.com/EBchecked/topic/591185/therapeutics. Accessed 12 March 2012

Fogel R (2004) Health, nutrition, and economic growth. Econ Dev Cult Change 52:643–658

Fuster V, Kelly B (eds) (2010) Promoting cardiovascular health in the developing world: a critical challenge to achieve global health. The National Academies Press, Washington

India Knowledge @ Wharton (2011) Healthy business: will medical tourism be India's next big industry? Wharton School, University of Pennsylvania, 2 June 2011

International Institute for Population Sciences and Macro International (2007) National family health survey (NFHS-3), 2005–06: India. http://www.measuredhscom/aboutsurveys/search/metadata.cfm?surv_id=264&ctry_id=57&SrvyTp=available. Accessed 24 Feb 2012

Kaiser Family Foundation (2011) The global HIV/AIDS epidemic. http://www.kff.org/hivaids/upload/3030-16.pdf. Accessed 5 Feb 2012

Khanna T et al (2005) Narayana Hrudayalaya heart hospital: cardiac care for the poor. Harvard Business Publishing, Boston

Knox R (2007) India's doctors returning home. National Public Radio, 30 November 2007

Markets and Markets (2011) Global biomaterials market: 2010–2015. http://www.marketsandmarkets.com/Market-Reports/biomaterials-393.html

Moussy F (2010) Biomaterials for the developing world. Wiley InterScience, New York

Ramakrishna S et al (2010) Biomaterials: a nano approach. CRC Press, New York

Ratner B et al (2004) Biomaterials science, 2nd edn. Elsevier Academic Press, San Diego

Rosenthal V et al (2006) Device-associated nosocomial infections in 55 intensive care units of 8 developing countries. Ann Intern Med 145:582–591

Subramanian S et al (2008) Health inequalities in India: the axes of stratification. Brown J World Affairs 14:127–138

UNICEF (2008) The state of the world's children 2008: child survival. United Nations Children's Fund, New York

UNICEF (2009) The state of the world's children 2009: maternal and newborn health. United Nations Children's Fund, New York

World Health Organization (2006) Country cooperation strategy: 2006–2011. http://www.who.int/countryfocus/cooperation_strategy/ccs_ind_en.pdf. Accessed 10 Feb 2012

World Health Organization (2008) Closing the gap in a generation: health equity through action on the social determinants of health. http://www.searo.who.int/LinkFiles/SDH_SDH_FinalReport.pdf. Accessed 1 March 2012

World Health Organization (2009) Disease and injury country estimates. http://www.whoint/healthinfo/global_burden_disease/estimates_country/en/index.html. Accessed 15 Feb 2012

World Health Organization (2010a) Global atlas of the health workforce. http://www.who.int/globalatlas/autologin/hrh_login.asp. Accessed 15 Feb 2012

World Health Organization (2010b) National health accounts (NHA). http://www.who.int/nha/country/en. Accessed 15 Feb 2012

World Health Organization (2011a) World health statistics. http://www.who.int/whosis/whostat/EN_WHS2011_Full.pdf. Accessed 2 Feb 2012

World Health Organization (2011b) Mortality data. http://www.who.int/healthinfo/statistics/mortality/en/. Accessed 3 Feb 2012

World Health Organization (2011c) Mortality estimates for WHO Member States in 2008. http://www.who.int/entity/healthinfo/statistics/bodgbddeathdaly.estimates.xls. Accessed 7 Feb 2012

Chapter 2
Natural Resources with Potential for Health Applications

If one simply drives through the rural areas or observes the rituals of one of India's dozens of harvest festivals, it is obvious that agriculture plays a pivotal role in Indian life. India is the second largest user of arable land in the world (after the United States) and agriculture accounts for approximately 16 % of India's GDP.[1] It is also important to note that despite the rising populations of urban hubs like Mumbai and New Delhi, more than three of every five Indians is dependent upon agriculture for his/her livelihood.[2] Silk farming, also known as sericulture, is also a major operation in India, employing approximately 750,000 people.[3]

Hence, this chapter will introduce nine natural resources in India, all of which are major factors in agriculture or sericulture. Each of these resources has also demonstrated potential for biomaterial and therapeutic applications, an area that will be explored further in Chap. 3.

Bamboo

One of the world's fastest-growing plants, bamboo is a versatile renewable resource with applications ranging from construction materials to culinary ingredients.

Bamboo, depicted in Fig. 2.1, is the everyday name for perennial, ornamental grasses of the family *Gramineae* that can be subdivided into five general with about 280 different species. Most bamboo development can be found in Asia's monsoon regions, but bamboo is also quite prevalent in the Americas and Africa. Bamboo plants typically appear as straight, upright stems emerging from rhizomes

---

[1] US Department of Agriculture (2011).
[2] Soundari (2011).
[3] Central Silk Board, Bangalore (2011).

V. Eswarappa and S. K. Bhatia, *Naturally Based Biomaterials and Therapeutics*, SpringerBriefs in Public Health, DOI: 10.1007/978-1-4614-5386-4_2, © The Author(s) 2013

Fig. 2.1 Bamboo stalks (*left*) and Bananas growing on a tree (*right*) [from Wikimedia Commons Users annieo76 and Pisang Rebus/Saba (Sap Caio), respectively]

and topped by horizontal branches. Most of these grasses are woody, but some are herbaceous or climbing. The largest types of bamboo can grow to be a stunning 120 ft in height and 1 foot in diameter. Bamboo seeds and shoots are often used as food while leaves can be used for livestock fodder. The stems are highly versatile and can be used to create bridges, cooking vessels, furniture, paper pulp, weapons, and other items.[4] Particularly in India and China, bamboo has also been used for medicinal purposes.[5]

Covering nearly 9 million hectares of land, Indian bamboo accounts for 36 % of the world's bamboo.[6] The country produces approximately 3 million tonnes of bamboo each year and has had much success and high yields in exports. In fact, the bamboo industry has demonstrated such potential that the country has plans to use the industry to increase rural employment and livelihoods by creating several million new jobs and lifting associated families out of poverty.[7]

Banana

The banana plant is a large, tropical flowering plant of the family *Musaceae*, commonly found in South and Southeast Asia as well as the Americas.[8] One of the most common banana types, *Musa sapientum*, is thought to have originated in Asia and is considered one of the earliest cultivated fruits. In a banana plant, a pseudostem made up of concentric leaves rises from the rhizome. At the center of these leaves, an inflorescence stalk develops on which fruits later grow, as shown in Fig. 2.1. This seedless, edible fruits grow in bunches and typically mature within

---

[4] Strausbaugh and Core (2008).

[5] Farrelly (1984).

[6] Rawat and Khanduri (1999).

[7] NABARD (2007).

[8] Lentfer and Boyd (2012).

75–150 days while growing to a length of approximately 6–8 in.[9] The fruits are easy to consume, low in sodium, and rich in calcium and potassium.[10] Although the fruit is commonly used for culinary purposes, the other parts of the banana plant can be used for such diverse functions as musical instruments or disposable dishware.

As of 2008, global banana production averaged at around 100 million bunches per year. The majority of global banana production takes place in the Western Hemisphere—in countries like Ecuador, Guatemala, and Honduras—and approximately three-fifths of bananas worldwide are ultimately consumed in the United States (see Footnote 9). India accounted for approximately 27 % of world banana production as of 2010 (see Footnote 10).

Coconut

Another common crop in India is coconut, also known as *Cocos nucifera*. The coconut is a large palm plant of the family *Areacaceae*, grown throughout tropical regions primarily for its fruit and fiber. The ovoid coconut fruit is typically 10 in. or greater in length and obtusely triangular in cross-section. A fibrous outer husk covers a spherical nut that consists of fleshy, oil- and protein-rich meat surrounded by a bony shell. Dried coconut meat, also called copra, is frequently sold for culinary uses, while the oil is often used for margarine, soap, or industrial purposes.[11] Ground coconut meat can also yield coconut milk, a common culinary ingredient. Coconut water is a popular drink taken from young coconuts that is rich in mineral ions, particularly potassium.[12] Some remaining parts of the coconut fruit can be used for animal feed. Coconut husks are used to produce coir fiber, which is made into ropes, matting, planting material, and upholstery filling (see Footnote 11).

Coconut is largely grown in Asia and Oceania, though it is also cultivated in the Americas.[13] In 2010, India alone produced 10,824,100 tonnes of coconut, making it the largest coconut producer in the world.[14,15] The crop contributes Rs. 70 billion to India's GDP every year.[16] As depicted in Fig. 2.2, coconut is an extremely versatile crop, used for everything from beauty products (such as coconut oil applied to the hair) to potting compost (like coir). Coconut plants can also be used for building material, thatching, apparel, dishware, baskets, and other purposes (see Footnote 11). Since every bit of the coconut can be used, the plant

---

[9] Schroeder and Dimitman (2008).

[10] Mohapatra et al. (2010).

[11] MacDaniels and Chiarappa (2008).

[12] Yong et al. (2009).

[13] Ibid

[14] FAOSTAT (2012).

[15] National Horticulture Board of India (2010).

[16] Markrose (2012).

Fig. 2.2 Different forms and uses of coconut. From *left* coconuts on a tree, hut made with thatched coconut leaves, coir fiber (from Wikimedia Commons Users Iaminfo, Sengai Podhuvan, and Fotokannan, respectively)

is sometimes called the "Tree of Life". During World War II and the Vietnam War, coconut water was even used as a substitute for intra-venous solutions when the latter was in short supply.[17]

Jackfruit

Jackfruit (species name *Artocarpus heterophyllus*) is a plant commonly found in regions of South and Southeast Asia, though it can also be found in places like Brazil where it is considered an invasive species. Archeological evidence indicates that jackfruit—a member of the *Moraceae* family—was cultivated in India over 3,000 years ago.[18] Jackfruit, which is related to the breadfruit, is primarily grown for its fruit and sturdy wood. When fully mature, jackfruit plants are typically 50–70 ft in height with stiff, glossy, green leaves between 6 and 8 in. in length. Their fruits, illustrated in Fig. 2.3, can weigh up to 40 pounds and grow to be about 2 ft long.[19] Therefore, they are considered the largest tree-borne fruits in the world.[20]

The unripe fruit, greenish in color, can be cooked as a vegetable while the brown ripened fruit is typically consumed fresh and distinguished by its "sweetly acid but insipid" pulp (see Footnote 19). The fruit is often used as a food or for herbal remedies, while the rest of the plant can be used for medical purposes, for construction, or to create musical instruments. In addition, the jackfruit plant can be used to provide shade for other crops or to yield an orange-hued dye which is often used to color the garments of Buddhist priests.[21]

---

[17] Library of Congress (2010).

[18] Jain et al. (2011).

[19] Encyclopaedia Britannica (2012).

[20] Valavi et al. (2011).

[21] Elevitch and Manner (2006).

Fig. 2.3 Jackfruits on a tree (*left*) and jute drying in the sun (*right*) (from Wikimedia Commons Users Ahoerstemeier and Auyon, respectively)

Jute

Jute is a vegetable fiber product of the family *Tiliaceae* often used to make coarse cloth, geotextiles, sacks, or twine.[22] Most of the world's largest jute producers are in Asia, with Bangladesh, China, India, and Myanmar contributing the lion's share.[23] However, jute is quite common in tropical Africa as well.[24]

Jute can be divided into two categories: White (*Corchorus capsularis*) and Tossa (*Corchorus olitorius*). The former generally has smaller, straight leaves that are dark green in color with deep serrations, while the latter typically has larger, drooping leaves of a yellow–green hue.[25] The plants are typically slender, half-shrubby annuals that reach 8–12 ft in height. Harvested stems are usually retted in pools of water to rot out softer tissue. Then the stems are beaten on the surface of the water to loosen the jute fiber strands. These fibers are then collected, dried in the sun (see Fig. 2.3), and prepared for use. Although jute fibers can weaken substantially in the presence of moisture, jute is used quite commonly for everyday purposes that can involve exposure to moisture, such as crop transportation (see Footnote 22).

India is the world's largest producer of raw jute and jute products and 4 million Indian families are involved in the jute cultivation process.[26] In 2010, India

---

[22] Nelson and Summers (2008).

[23] International Jute Study Group (2003a).

[24] Oboh et al. (2009).

[25] Indian Council of Agricultural Research (2010).

[26] National Multi-Commodity Exchange of India (2012).

Fig. 2.4 Rice field, *Bombyx mori* Larvae, and *Bombyx mori* Cocoons (*left* to *right*) (from Wikimedia Commons Users Miya, Małgorzata Miłaszewska, and Katpatuka, respectively)

produced over 1.7 million tonnes of jute (see Footnote 14). Annual worldwide production of jute and allied fibers is approximately 3 million tonnes.[27]

Rice

Oryza sativa, which is commonly referred to as rice, is a tremendously important cereal crop that serves as the staple food for more than half of the world's population.[28] In some Asian countries, rice is a more crucial carbohydrate source than wheat, with yearly per capita rice consumption reaching levels of up to 50 times that of the United States. Rice is thought to have originated around 7 millennia ago in areas of present-day China and India, though it is a common crop throughout the Americas, Africa, Asia, and Oceania today.[29] India itself produced over 120 million tonnes of rice paddy in 2010 (see Footnote 14).

Rice is unique in that it can grow in very wet environments that many other crops would not tolerate (see Footnote 28). This is shown in Fig. 2.4, where rice is growing in a field submerged in water. The rice plant itself is typically between 2 and 6 ft in height, and new shoots will grow given sufficient space and soil fertility. These annual plants can take 80–200 days to mature. A rice crop can be harvested by hand or using large combines, similar to what is often used for wheat (see Footnote 29).

More than 95 % of the global rice crop is used as a human food source. The rice kernel consists of four parts: hull/husk, seedcoat/bran, embryo/germ, and endosperm (see Footnote 29). Rice is extremely rich in complex carbohydrates—an excellent energy source—although many nutrients and minerals are lost during the milling and polishing process that removes the outer rice husk and bran to turn

---

[27] International Jute Study Group (2003b).

[28] International Rice Research Institute (2012).

[29] Rutger et al. (2008).

brown rice into white rice (see Footnote 28). Rice can also be used to produce alcoholic beverages, livestock feed, and fuel sources. Rice straw can be used for thatching, weaving ropes or bags, or making baskets, hats, mats, and paper (see Footnote 29).

Silk

Although it can be produced from various sources, silk generally refers to the glossy fiber derived from silkworm larvae. Silkworms have been raised specifically for silk production purposes for at least 5 millennia in a process termed sericulture. The *Bombyx mori*, one of the most important and common silkworms in the world, spins a peanut-shaped cocoon during the second stage of its life cycle as a means of protection and to enable its transition to a pupa. The *Bombyx mori* larvae and cocoons are illustrated in Fig. 2.4. The silkworm secretes a protein-based silk film and usually finishes its cocoon within 3 days, at which point its life cycle can be terminated to allow for the silk to be harvested and unwound. This silk is typically composed of 95 % protein, of which about four-fifths is fibroin and one-fifth is sericin. The fiber is a double-filament of fibroin held together by the gummy sericin.[30]

As the world's largest silk producer after China, India accounted for nearly 15% of the 140,051 metric tonnes that comprise the world's total raw silk production in 2010. It is also the only country to produce all four types of silk: mulberry, tasar, eri, and muga. Silk consumption in India itself is around 29,300 metric tonnes per year, but there is high export demand as well, particularly to regions such as the US and Europe (see Footnote 3). The US is the world's greatest importer and consumer of silk (see Footnote 30). Nearly three-quarters of a million people are employed in sericulture in India, many of whom are classified as "disadvantaged" or women (see Footnote 3).

Soy

Glycine max, also known as the soybean, is a legume that has been grown in Asia for over 3,000 years and is currently produced largely in Argentina, Brazil, China, India, and the United States.[31] India alone produced nearly 10 million tonnes of soybean in 2010 (see Footnote 14).

The annual soybean plant can take 90–180 days to reach maturity, at which point it is typically between 25 and 50 in. in height. Two unifoliolate leaves develop above

---

[30] Potter (2008).
[31] Soyatech (2012).

Fig. 2.5 Soybeans on plant, tamarind pods and components, and tamarind fruits on a tree (*left* to *right*) (from Wikimedia Commons Users Scott Bauer, Carschten, and B.navez, respectively)

the cotyledon while all other leaves are typically trifoliolate. Depending on space available, a plant may have a few branches (which emerge from the main stem) with just a few seeds or have many branches with a thousand or so seeds. Soybean plants also have white or purplish flowers, and much of the plant is covered with thin brown or gray hairs. There are thousands of varieties of soybean, but nearly all soybean pods contain one to four seeds with black, brown, green, or yellow seed coats, as shown in Fig. 2.5. The seeds are considered the most important part of the plant, and contain approximately 40 % protein and 20 % oil on a dry-weight basis.[32] Genetically modified soybeans have been commercially grown since 1996 and are now dominant in most major producing countries (see Footnote 31).

Soy processing generally converts soybeans into oil, meal, or protein products. The three major processing steps are oil extraction, oil refining, and protein processing (see Footnote 32). About 85 % of the soybeans produced in the world are processed into soybean meal and oil. Nearly all of the soybean meal is further processed into animal feed and fertilizer, while a smaller fraction is used for soy flour and proteins. About 95 % of the oil fraction is used as edible oil in items like salad dressing or mayonnaise, with the rest going to industrial products like biodiesels, soaps, or varnishes (see Footnote 31).

Tamarind

Although the tamarind, of family *Leguminosae*, is considered native to tropical Africa, it may also be indigenous to India. Today, it is widely grown elsewhere in Asia and in Mexico as well.[33] The *Tamarindus indica* is an evergreen tree that can grow to be nearly 80 ft in height and over 10 ft in girth. The plant grows pale yellow or pink flowers and podded fruit. The subcylindrical pod is generally 4–7 in. in length and about 1.5 in. in width with a velvety brown cover, as depicted

---

[32] Fehr et al. (2008).

[33] California Rare Fruit Growers, Inc. (1996).

in Fig. 2.5. The pod's shell is brittle, but inside are 3–10 hard seeds about 0.6 in. in length, covered in a sticky, edible pulp.

Tamarind fruits have one of the highest levels of carbohydrate (41.1–61.4 g/100 g) and protein (2–3 g/100 g) of any known fruit.[34] These podded fruit are primarily used for medicinal or culinary purposes, though other parts of the plant have different uses. For example, tamarind oil can be used for varnish and tamarind wood is used to create everything from canoes to printing blocks.[35]

India is Asia's largest sour tamarind producer.[36] The country exported 37,000 metric tonnes of tamarind from 2001 to 2002.[37]

References

Bhadoriya S et al (2010) *Tamarindus indica:* extent of explored potential. Pharm Rev 5:14

California Rare Fruit Growers, Inc. (1996) Tamarind. http://www.crfg.org/pubs/ff/tamarind.html. Accessed 7 Febr 2012

Central Silk Board, Bangalore (2011) Note on the performance of Indian Silk Industry & functioning of Central Silk Board, 14 Dec 2011. http://www.csb.gov.in/assets/Uploads/pdf-files/Note-on-SERI.pdf. Accessed 11 Febr 2012

Elevitch C, Manner H (2006) *Artocarpus heterophyllus* (jackfruit). In: Craig R. Elevitch (ed) Species profiles for Pacific Island agroforestry, Permanent Agriculture Resources, Holualoa

El-Siddig K et al (2006) Tamarind, Tamarindus indica. Southampton Centre for Underutilised Crops, Southampton, UK

Encyclopaedia Britannica (2012) Jackfruit. Encyclopaedia Britannica Online Academic Edition. http://www.brittanica.com/EBchecked/topic/298742/jackfruit. Accessed 13 March 2012

FAOSTAT (2012) Statistics. Food and Agriculture Organization of the United Nations. http://www.faostat.fao.org/site/567/DesktopDefault.aspx?PageID=567#ancor. Accessed 12 Febr 2012

Farrelly D (1984) The book of bamboo. Sierra Club Books, San Francisco

Fehr W et al (2008) Soybean. AccessScience, McGraw-Hill Companies, New York. http://www.accessscience.com. Accessed 16 March 2012

Indian Council of Agricultural Research (2010) What is the difference between tossa and white jute? http://www.icar.org.in/node/486. Accessed 17 Febr 2012

International Jute Study Group (2003) Growing regions. http://www.jute.org/growing_regions.htm. Accessed 14 March 2012

International Jute Study Group (2003) Jute, Kenaf & Roselle plants. http://www.jute.org/plant.htm. Accessed 14 March 2012

International Rice Research Institute (2012) Rice basics. http://www.irri.org/about-rice/rice-facts/rice-basics. Accessed 15 Febr 2012

Jain S et al (2011) Anti-inflammatory activity of *Artocarpus heterophyllus* bark. Der Pharmacia Sinica 2:127–130

Janick J, Paull RE (eds) (2008) Tamarindus. In: The encyclopedia of fruit & nuts. Cambridge University Press, Cambridge

Kaur G et al (2006) Tamarind: date of India. Sci Tech Entrepreneur, Dec 2006

---

[34] El-Siddig et al. (2006).

[35] Bhadoriya et al. (2010).

[36] Janick and Paull (2008).

[37] Kaur et al. (2006).

Lentfer C, Boyd W (2012) Tracing antiquity of banana cultivation in Papua New Guinea. The Australia & Pacific Science Foundation. http://www.apscience.org.au/projects/PBF_02_3/pbf_02_3.htm. Accessed 20 Febr 2012

Library of Congress (2010) Interesting coconut facts, 23 August 2010. http://www.loc.gov/rr/scitech/mysteries/coconut.html. Accessed 29 Febr 2012

MacDaniels L, Chiarappa L (2008) Coconut. AccessScience, McGraw-Hill Companies, New York. http://www.accessscience.com. Accessed 16 March 2012

Markrose V (2012) Coconuts in India. Botanic Gardens Conservation International. http://www.bgci.org/education/1685/. Accessed 20 Febr 2012

Mohapatra D et al (2010) Banana and its by-product utilization: an overview. J Sci Ind Res 69:323–329

NABARD (2007) Bamboo cultivation. National Bank for Agriculture and Rural Development. http://www.nabard.org/modelbankprojects/forestry_bamboo.asp. Accessed 10 Febr 2012

National Horticulture Board of India (2010) Commodity bulletin. http://www.nhb.gov.in/statistics/commodity-bulletin.html. Accessed 26 Febr 2012

National Multi-Commodity Exchange of India (2012) Report on jute. http://www.nmce.com/files/study/rawjute.pdf. Accessed 7 Febr 2012

Nelson E, Summers T (2008) Jute. AccessScience, McGraw-Hill Companies, New York. http://www.accessscience.com. Accessed 17 March 2012

Oboh G et al (2009) Characterization of the antioxidant properties of hydrophilic and lipohilic extracts of Jute (*Corchorus olitorius*) leaf. Int J Food Sci Nutr 60:124–134

Potter M (2008) Silk. AccessScience, McGraw-Hill Companies, New York. http://www.accessscience.com. Accessed 16 March 2012

Rawat J, Khanduri D (1999) The status of bamboo and rattan in India. International Network for Bamboo and Rattan

Rutger J et al (2008) Rice. AccessScience, McGraw-Hill Companies, New York. http://www.accessscience.com. Accessed 17 March 2012

Schroeder C, Dimitman J (2008) Banana. AccessScience, McGraw-Hill Companies, New York. http://www.accessscience.com. Accessed 16 March 2012

Soundari MH (2011) Indian agriculture and information and communications technology. New Century Publications, New Delhi

Soyatech (2012) Soy Facts. http://www.soyatech.com/soy_facts.htm. Accessed 11 Febr 2012

Strausbaugh P, Core E (2008) Bamboo. AccessScience, McGraw-Hill Companies, New York. http://www.accessscience.com. Accessed 17 March 2012

US Department of Agriculture (2011) India: basic information. Economic Research Service, Briefing Rooms, 18 August 2011. http://www.ers.usda.gov/briefing/india/basicinformation.htm. Accessed 4 Febr 2012

Valavi S et al (ed) (2011) The Jackfruit. Studium Press LLC, Houston

Yong J et al (2009) The chemical composition and biological properties of coconut (*Cocos nucifera* L.) water. Molecules 14:5144–5164

Part II

Chapter 3
Biomaterial and Therapeutic Applications

Bamboo Applications

Blood Purification

Sepsis, a condition in which the bloodstream is fighting a systemic infection, is a major health concern in both developed and developing countries. As of 2006, severe sepsis claimed the lives of 1500 people worldwide every day.[1] Sepsis is also considered to be the largest cause of global neonatal mortality, responsible 30–50 % of all neonatal deaths in developing countries. In a study conducted by the National Neonatal Perinatal Database from 2002 to 2003, neonatal sepsis occurred in 3 of every 100 live births in India.[2] Sepsis can be caused by bacteria or other sources and includes illnesses such as meningitis, pneumonia, and septicemia.[3] Hepatic (liver) failure is another major cause of sepsis, and hemoperfusion—a process in which the blood is filtered by passage through a cartridge containing adsorbent particles such as activated charcoal—is a standard way to remove toxins from the blood of patients suffering from this condition.[4]

Research has demonstrated that bamboo can be used to create effective blood purification agents for hemoperfusion. *Phyllostachys pubescens*, also known as Moso bamboo, grows predominantly in China but is present in India as well. After preparation by carbonization and steam activation, bamboo charcoal particles can be formed and bound together with biocompatible chitosan to form bamboo charcoal beads (BCBs). Studies of cell viability, hemolysis, and cytokine secretion from macrophage cells demonstrated the BCBs' low toxicity levels and the weak

---

[1] Hsieh et al. (2010).
[2] Sankar et al. (2008).
[3] Mayo Clinic Staff (2012).
[4] Rahman et al. (2006).

V. Eswarappa and S. K. Bhatia, *Naturally Based Biomaterials and Therapeutics*, SpringerBriefs in Public Health, DOI: 10.1007/978-1-4614-5386-4_3, © The Author(s) 2013

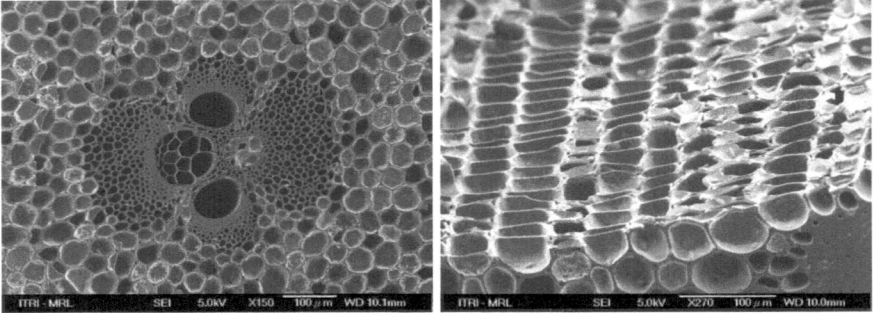

Fig. 3.1 Bamboo charcoal surface morphology, two views via SEM micrograph (Hsieh et al. (2007)

immune response it elicited.[5] Two views of the bamboo charcoal surface morphology are shown in Fig. 3.1.

Activated carbon is commonly used to remove particular toxins from blood, but the degradation of tiny carbonaceous particulates can endanger patients' health by obstructing or getting stuck in blood vessels. In addition, current methods run the risk of using activated carbon contaminated with heavy metal from fossil fuels.[6, 7] However, BCBs biodegrade and do not pose the same contamination threat. Not only are these bamboo-based purifying agents biocompatible, they are low-cost as well, increasing their suitability to the Indian medical environment.

Bone Repair or Replacement

With about 1 in 10 deaths in India being caused by injuries, it is no wonder that trauma and orthopedic applications have become crucial to improving the country's health outcomes.[8] Bioceramic, metallic, and polymeric biomaterials have all been used to try to address these needs. However, they all face challenges in their application. Bioceramic implants typically exhibit poor resistance against fatigue failure and low fracture toughness, which poses a problem for bone-repairing purposes. Due to their elasticity mismatch with bone, metallic materials will lead to stress-shielding and bone resorption. Polymeric materials possess a low modulus of elasticity, which limits their applications as well.[9]

---

[5] Hsieh et al. (2010).

[6] Chandy and Sharma (1993).

[7] Chandy and Sharma (1998).

[8] Joshipura (2008).

[9] Li et al. (1996).

As a result, bamboo has emerged as a promising biomaterial for bone repair and replacement applications. Researchers have found that there are many structural similarities between bamboo and human long bones. For example, bamboo has a gradient structure in the radial direction resembling the bone's gradient structure from outer cortical to inner spongy bone, indicating the potential application of bamboo for bone fracture fixation. In addition, across the bamboo stem (for the bamboo *Phyllostachys bambusoides*), the average modulus elasticity is 18 GPa, which is the same as the value for human cortical bone. Bamboo also has fairly high longitudinal tensile, bending, and compressive strengths, all of which are necessary for a good biomaterial meant to repair or replace a load-bearing bone.[10] The fact that bamboo also possesses silicon on the inner and outer surfaces of its stem means that it may have an easier time conducting bone mineralization.[11] Plus, studies have shown that the porous nature of bamboo can help with the ingrowth and anchorage of bone.[12]

One study demonstrated that, owing to its silica components, bamboo is able to induce precipitation of calcium phosphate *in vitro*. This is a particularly significant development because many scientists believe that the formation of such an apatite layer is necessary for a biomaterial to properly bond to bone. After chemically removing surface fatty substances, the researchers grafted the bamboo with a polyethylene glycol-based polymer that can facilitate calcification because its polyether soft segment polymer can cause metal-ion chelation. The grafted bamboo samples were then soaked in one of two calcification solutions: accelerated calcification solution (ACS) or simulated body fluid (SBF). The bamboo then formed continuous layers of calcium phosphate, a major component of human bone. Although the bamboo did possess some cytotoxic leachable components that could pose a threat to patients, they could be largely removed by ethanol, methanol, and toluene.[13]

In another study, charcoal bamboo (*Bambusae glaucescens*) was specially fired, sterilized, and shaped before being inserted in the resected tibial diaphysis of white rabbits. At 6 weeks, there was neobone invasion into the bamboo's pores, and good bony ingrowth was seen months later. Another promising finding was that there was no evidence of fibro-encapsulation or inflammatory response in the bone marrow or at the area where the bone and bamboo made contact.[14] Therefore, charcoal bamboo was found to be a suitable alternative bone substitute, and future studies should explore potential applications in humans.

---

[10] Li et al. (1996).

[11] Charnot and Peres (1978).

[12] Bednar et al. (1982).

[13] Li et al. (1997).

[14] Kosuwon et al. (1994).

Neuroprotector

Neurological disorders are becoming an increasingly significant problem in Indian health. Based on prevalence studies of common neurological disorders conducted among rural and urban populations in the Bangalore region, some researchers have estimated that between 20 and 30 million Indians suffer from neurological disorders of various sorts. One study conducted in 1987 and replicated in 2004 revealed a nearly five-fold increase in the incidence of Parkinson's disease over that time period, from 0.07 to 0.33 per 1,000 people.[15]

Bamboo has demonstrated the potential to protect against neuronal damage and neurodegenerative disorders—such as Alzheimer's, Parkinson's, and Huntington's diseases—where inflammation and oxidative stress are implicated. A relatively recent development in the effort to combat such diseases is the strategy of attenuating the transduction of apoptotic signals in order to prevent neuronal cell death.[16] Studies have shown that one way to go about doing this is to use lignin, a durable network polymer that accounts for 20–30 % of global plant biomass, making it the second most abundant organic plant substance.[17] Scientists have devised a system to convert lignin into derivatives including lignophenols, which are very active, stable, and phenolic. These lignophenols demonstrated impressive neuro-protective abilities against oxidative stress-induced apoptosis. In one study, lig-8—a lignocresol derived from bamboo—derived the strongest neuro-protective activity against hydrogen peroxide-induced apoptosis in human neuroblastoma cells. Lig-8 also showed an anti-apoptotic effect by impeding the dissipation of the mitochondrial membrane permeability transition. In addition, flow cytometry revealed that lig-8 exhibited antioxidant properties in the cells exposed to hydrogen peroxide. Therefore, bamboo-derived lig-8 has the potential to be an effective neuro-protector, altering the signaling pathway of neuronal cell death and slowing the progress of neurodegenerative diseases.[18]

Another study has demonstrated the potential of lig-8 to protect against neuronal damage caused by several other sources: oxygen-glucose deprivation followed by re-oxygenation, tunicamycin, and proteasome inhibitor. The study examined the activity of lig-8 in two scenarios of oxygen-glucose deprivation (OGD) stress-induced neurotoxicity, which may be related to endoplasmic reticulum stress. The first was in the common neuronal differentiation model of pheochromocytoma (PC12) cells while the second was in vivo against retinal damage-induced mice. The tunicamycin—a glycosylation inhibitor that is cytotoxic to many cells[19]—induces cell death in PC12 cells. The proteasome inhibitor (PSI) also induces cell death, but in human neuroblastoma cells. Lig-8

---

[15]   Gourie-Devi (2008).

[16]   Akao et al. (2004).

[17]   Ito et al. (2006).

[18]   Akao et al. (2004).

[19]   Varki et al. (1999).

demonstrated impressive neuro-protective effects in each case. OGD-induced cell damage displayed an inverse relationship to the concentration of lig-8 applied, with lig-8 at a 30 μM concentration demonstrating the same level of cell viability as the control PC12 cells. Lig-8 also inhibited tunicamycin-induced cell death in PC12 cultures, with the greatest neuro-protective effect occurring at a concentration of approximately 3 μM. Lig-8 treatment also substantially decreased PSI-induced apoptotic cell death in SH-SY5Y cells.

It is likely that lig-8's neuro-protective abilities are related to lignin's physiological effects that enable woody plants to live longer than many animals. Lignin in such plants may physically protect plant cells by acting as a glue-like substance while also offering some sort of biological protection. Such potential behavior can be further explored in future studies. Regardless of the outcome of such studies, since the bioactive lignophenols derivative lig-8 demonstrated neuro-protective effects in both in vitro and in vivo models, it is likely that it will also exhibit such abilities when it has penetrated living human tissue.[20]

Banana Applications

Burn Dressing

As of 2004, burns were the 8th largest cause of death worldwide in people aged 15–29 years and 10th largest in people aged 5–14 years.[21] Today, fire-related burns disfigure and disable millions every year while killing hundreds of thousands of others.[22] Low- and middle-income countries bear a disproportionate amount of this burden, accounting for 95 % of fatal fire-related burns.[23] The situation is particularly stark in India, where the estimated annual burn incidence is somewhere between 6 and 7 million.[24] It is also likely that more than 100,000 Indians lose their lives to fire-related causes each year.[25] Currently, treating burns can be an expensive process involving hospital stays, costly medications, multiple operations, and extended rehabilitation periods. However, the vast majority of burn victims in India have low incomes, with an average per capita monthly income equivalent to less than $5 a month.[26] This poses an obvious affordability hurdle and has precipitated the search for a cheap but effective dressing alternative.

---

[20] Ito et al. (2006).

[21] World Health Organization (2010).

[22] Chandran (2010).

[23] Ebel et al. (2010).

[24] Gupta et al. (2010).

[25] Sanghavi (2009).

[26] Gore and Akolekar (2003).

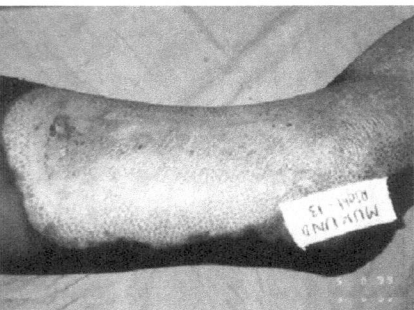

Fig. 3.2 Prepared banana leaf dressing above boiled potato peel bandage (*left*) and BLD applied to right leg of patient, 13 days post-burn (*right*). Gore and Akolekar (2003)

One Mumbai hospital used boiled potato peel bandages (BPPB) as an alternative and patients generally preferred them to the original Vaseline gauze dressings. However, there was still some discomfort associated with these bandages and their preparation was too time-consuming and challenging to teach and learn. In addition, although the BPPBs cost less than $1, they were still out of many patients' price range.[27]

Large, inexpensive, and easily attainable, the leaves of banana plants were found to address many of these problems. The leaves were cut down their midribs and pasted onto bandage cloth with a flour-based paste before being autoclaved for use in patients along with a topical agent. These banana leaf dressings (BLD) were compared to BPPBs in a 30 patient trial and found to be generally similar or superior. There were no significant differences between the two materials with respect to days taken for epithelialization, eschar formation, the need for skin to graft over deep partial thickness burns, or microbe growth. Images of the banana leaf dressings are shown in Fig. 3.2. Since minimal discomfort was experienced and the BLD was easier to prepare and 1/11th the cost of the BPPB, this banana-based biomaterial proved to be a promising development in the field of burn dressings (See foot note 27).

Inhibition of Viral Transmission

As of 2010, approximately 34 million people were living with HIV—over 95 % of whom resided in low- or middle-income countries—and there were over 7,000 new HIV infections every day. AIDS-related causes have claimed the lives of almost 30 million people already, and the virus has deleteriously affected the lives

---

[27] Gore and Akolekar (2003).

of countless others.[28] This global HIV epidemic has not overlooked India. Current HIV estimates in India range from 1.5 to 3 million infected people.[29, 30]

In addition, even though over 25 anti-HIV drugs were approved and availability of antiretroviral (ARV) therapies had been improving in low- and middle-income countries, by 2008, the rate of new HIV-1 infections was 2.5 times greater than the rate of new individuals receiving ARV drugs.[31] An effective HIV vaccine may be many years away and condoms—though useful in combating the spread of HIV—are often used incorrectly or inconsistently, particularly in areas where women have less control over their sexual activity. Therefore, the development of intravaginal or intrarectal microbicides may be a good alternative.[32] In fact, some predict that over just 3 years, covering one-fifth of the population with a 60 % effective microbicide could prevent up to 2.5 million new HIV infections.[33]

Several promising HIV transmission inhibitors block the virus before it integrates its genome into the target cell. Lectins—proteins that can recognize and bind carbohydrates without altering them—are able to inhibit the HIV-1 entry step by binding to the viral envelope's carbohydrate structures.[34] One HIV-1 envelope protein, called gp120, is an attractive target because it contains multiple sites for glycosylation, a process crucial to the viral life cycle. Studies have shown that banana lectin (BanLec) is able to bind to high mannose structures on gp120 to prevent the cellular attachment and entry of HIV. Cells treated with BanLec demonstrated lower levels of strong-stop DNA, an early HIV reverse transcription product that can often be detected between when the virus enters the cell and when the virus is uncoated. In addition, although most of banana lectin's HIV inhibitory activity derives from its ability to block viral attachment, some HIV inhibition was still observed when BanLec was added after attachment.[35] Therefore, BanLec could also be useful in preventing HIV transmission through post-attachment steps including virus fusion to the cell.

Using BanLec as an anti-HIV agent could also be effective because lectins can hinder the development of resistance by targeting multiple different glycosylation sites on the virus. Research has shown that for resistance to develop, many more mutations in the envelope sequence are needed than are typical. Administering a combination of different lectins would further reduce the chances that a lectin-based anti-viral therapy would fail due to resistance. Even if the virus developed a resistance to BanLec and other lectins, it is possible that this would increase its vulnerability to the normal human immune response. This is because the changes

---

[28] Kaiser Family Foundation (2011).

[29] United Nations Development Project, India (2011).

[30] Jha et al (2010).

[31] Joint United Nations Programme On HIV/AIDS (2008).

[32] Swanson et al. (2010).

[33] Watts (2002).

[34] Meagher et al. (2005).

[35] Swanson et al. (2010).

in glycosylation that allow the virus to develop resistance to lectins also increase its susceptibility to neutralizing antibody responses.[36]

Studies have shown that banana lectin microbicides have demonstrated the capacity to be as powerful as two major microbicides that have been developed. However, the BanLec treatments are less expensive and do not demonstrate the same side effects associated with those two.[37] Although there are still some safety concerns that need to be explored with BanLec, attaching recombinant therapeutic proteins to the lectin's polymer chains have been shown to reduce toxicity and alter bioavailability. Banana lectin also has potential antiviral applications beyond HIV. Glycosylation is a process crucial to other viruses as well, not just HIV-1. Particular lectins have already demonstrated inhibitory behavior with respect to viruses such as Ebola, hepatitis C, influenza, Marburg, and SARS coronavirus. BanLec could possibly be applied to combat the spread of such viruses as well.[38]

Therefore, banana lectin has exhibited potential to inhibit the transmission of viruses, particularly HIV. Bio-derived applications of this and other lectins should be explored to develop methods to better combat the spread of infectious disease, such as intravaginal microbicides to reduce HIV transmission.

Diabetes Treatment

Over the past few decades, the diabetes predicament in India has gotten progressively worse. As of 2007, India was the "diabetes capital of the world" with 41 million Indians having diabetes. This meant that one in five diabetics worldwide was Indian.[39] Predictions for the future do not paint a more optimistic picture, with an estimated 87 million Indians being diabetic by 2030.[40] Hence, there have been massive efforts to find effective treatments for diabetic patients.

The pseudostem and flower of banana plants, as pictured in Fig. 3.3, have been found to be effective in the treatment of diabetic rats. When consumed at a 5 % level in the diet, symptoms such as body weight, glomerular filtration rate, polydipsia, polyphagia, polyuria, and urine sugar were all ameliorated.[41] The methanolic extract of one type of banana fruit (*Musa paradisiaca*) has a hypoglycemic effect on diabetic mice[42] while pectin from a different type (*Musa sapientum*) had

---

[36] Swanson et al. (2010).

[37] Reed (2010).

[38] Swanson et al. (2010).

[39] Joshi and Parikh (2007).

[40] Joshi et al. (2012).

[41] Bhaskar et al. (2011).

[42] Ojewole and Adewunmi (2003).

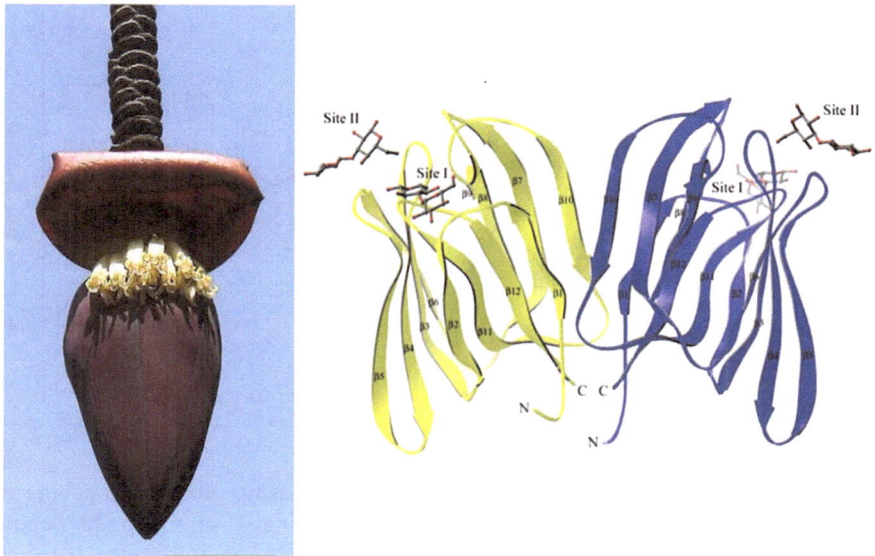

Fig. 3.3 Banana flower and chemical structure of BanLec. From Wikimedia Commons User Jim Conrad and Meagher et al. (2005)

a hypoglycemic effect on alloxan diabetic rats.[43] Hence, banana-derived compounds have demonstrated anti-diabetic effects that should be further investigated to examine potential for therapeutic applications.

Bone Grafting

Ligno-cellulosic banana fibers have also demonstrated potential for use in bone grafting substitutes. Banana fiber composites externally coated with calcium phosphate and hydroxyapatite could be useful for the fixation of fractured human bones, both internally and externally. Their biocompatibility combined with their strong mechanical properties could make banana fibers an ideal candidate for bone repair and replacement. In comparison to sisal and roselle fibers, banana fibers have a higher density and modulus of elasticity. Banana fibers are an especially appealing candidate given their current treatment as a waste material, so their incorporation in biomedical applications would put them to good use.[44]

---

[43] Pari and Umamaheshwari (2000).

[44] Chandramohan and Marimuthu (2011).

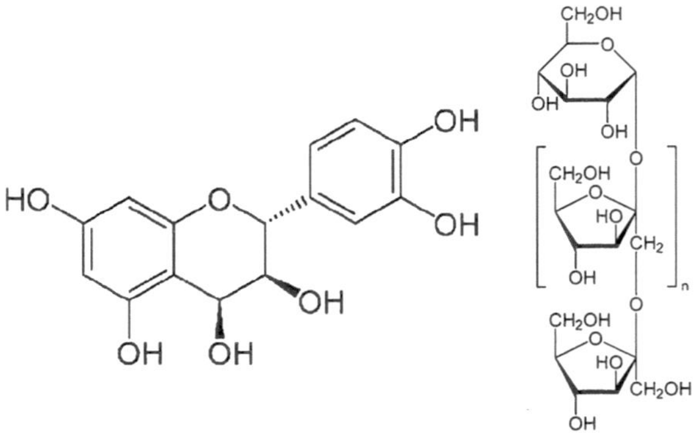

Fig. 3.4 Chemical structures of Leucocyanidin (*left*) and Inulin (*right*) (From Wikimedia Commons Users Nono64 and NEUROtiker respectively)

Anti-Ulcerogenic

Gastric ulcers are a painful condition posing an increasing problem to people in both the developing and developed worlds. Affecting 1 of every 20 people worldwide, gastric ulcers are thought to be caused by an imbalance between luminal acid and pepsin on one side and the mucosal layer, phospholipid layer, prostaglandins, and other factors on the other side. Many researchers are currently looking for more effective ways to treat gastric ulcers, which can be caused by such diverse factors as stress, nutritional deficiencies, and nonsteroidal-anti-inflammatory drugs.[45]

Multiple studies have demonstrated the protective effects that unripe plantain banana extract can have on gastric mucosa at risk for aspirin-induced damage. Therefore, a study was undertaken to examine the anti-ulcerogenic properties of the unripe plantain banana extract and its potentially active ingredient, the flavonoid leucocyanidin. The chemical structure of leucocyanidin is shown in Fig. 3.4. Active banana powder was prepared by drying fresh pulp at temperatures below 40 °C. This study showed that an aqueous, leucocyanidin-rich extract exhibited substantial anti-ulcerogenic properties, even when exposed to artificial gastric juice. The anti-ulcerogenic properties of such flavonoids have been partially explained by their reduction of acid secretion by gastric parietal cells.[46] This demonstrates the anti-ulcerogenic potential of banana-based compounds and further exploration of the topic could yield important therapeutic applications for human ulcer treatment.

---

[45] Rony et al. (2011).

[46] Lewis et al. (1999).

Iron Absorption Promoter

As one of the most common forms of malnutrition in the world, iron deficiency is a major global health concern that affected over 2 billion people at the end of the 20th century. Iron-deficiency anemia is a particularly significant problem for women in developing countries like India, where it is a common cause of maternal morbidity and mortality, pregnancy and delivery complications, and child-development problems.[47] In 1990, the Government of India claimed that nearly 1 in every 5 maternal deaths in the country was related to anemia, and 2 years later the World Health Organization estimated that a staggering 88 % of pregnant Indian women were anemic.[48] What is perhaps even more troubling is that despite widespread efforts and programs to lower the levels of anemia in India, the prevalence of anemia in pregnant mothers actually increased from 49.7 % in 1999 to 58.7 % in 2006.[49] This indicates a need for a new approach to anemia treatment.

One promising approach involves the use of inulin, a fructan, from bananas. The chemical structure of inulin can be found in Fig. 3.4. This type of inulin is associated with decreased sulfide concentrations and increased concentrations of soluble iron in the colon digesta of pigs. In addition, the increase in iron absorption was not accompanied by a change in the digesta's pH or phytase activity, which could have negative effects on other digestive and absorptive properties. Since pigs have a similar anatomy and physiology of the gastrointestinal tract as humans and their iron status can be easily manipulated by iron injection dosages, these animals are a good model through which to study human iron absorption.[50]

Studies have also shown that inulin enhances iron absorption and retention as well as hemoglobin repletion efficiency and hemoglobin recovery from anemia in rats.[51] Therefore, banana inulin-based therapeutics have good potential to reduce iron deficiency in humans and lower the burden of anemia in India.

Coconut Applications

Obesity and Diabetes Treatment

Studies indicate that hyperinsulinemia—a condition in which blood insulin levels are higher than normal for non-diabetics—is closely related to particular pathophysiological changes that occur during the progress of type 2 diabetes,

---

[47] Government of India. "Chapter 7: Nutrition and the Prevalence of Anaemia." From National Family Health Survey-2.

[48] Bentley and Parekh (1998).

[49] Klemm et al. (2011).

[50] Yasuda (2007).

[51] Ibid.

cardiovascular disease, and cancer.[52] High glucose levels in the blood can also lead to additional glycation products which might create unwanted functional and morphological alterations.[53] Since postprandial (after-meal) hyperglycemia in type 2 diabetic patients can lead to complications and diets with high glycemic indices can boost one's chances of developing cardiovascular disease, insulin resistance, and type 2 diabetes, it has become increasingly important to adequately control blood glucose and insulin response levels to avoid disease development. This has become a substantial problem with the increasing percentage of people who are overweight or obese from excess dietary energy intake, which can induce insulin resistance.[54]

Coconut is a rich source of D-xylose, a monosaccharide that is thought to suppress the postprandial glucose surge. This could be a very important tool in treatments to combat obesity-related diseases, an area where postprandial hyperglycemia is believed to play a significant role.[55] This sucrase inhibitor has been shown to lower intestinal sucrose activity by over 50 % in rats.[56] D-xylose in 5 g and 7.5 g doses has also lessened increases in serum glucose and insulin levels and slowed down glucose absorption in a study of 49 humans.

In one experiment, a control group consumed 50 g sucrose while a test group consumed 5 or 7.5 g xylose in addition to 50 g sucrose. Results showed that serum glucose levels after 30 min were substantially lower among those subjects who were administered xylose. Similarly, after 15 min, mean serum insulin was far lower in the test group. All of this indicates that coconut-derived D-xylose can be used to positively alter the postprandial glucose and insulin response and successfully inhibit the effects of obesity- and diabetes-related conditions.[57]

With respect to diabetes specifically, coconut has also demonstrated significant anti-diabetic effects. Research has shown that the hydro-methanol extract of *Cocos nucifera* (HECN) exhibited an anti-hyperglycemic effect in streptozotocin (STZ) induced diabetic rats that was comparable to that of gilbenclamide, a common anti-diabetic drug. It is thought that HECN's glucose-lowering abilities might be due to an increase in pancreatic insulin secretion from existing beta cells or release from bound insulin.[58]

The diabetic rats demonstrated elevated levels of triglyceride (TG) and cholesterol along with decreased HL levels. This could be due to an insulin shortage that activated lipase enzymes to hydrolyze stored TG and release fatty acids and glycerol into the blood. Increased TG and cholesterol in the rats' blood could result in cardiovascular disease. However, treatment with HECN improved the rats' lipid

---

[52] Bae et al. (2011).

[53] Gavin (2001).

[54] Bae et al. (2011).

[55] Ibid.

[56] Seri et al. (1996).

[57] Bae et al. (2011).

[58] Naskar (2011).

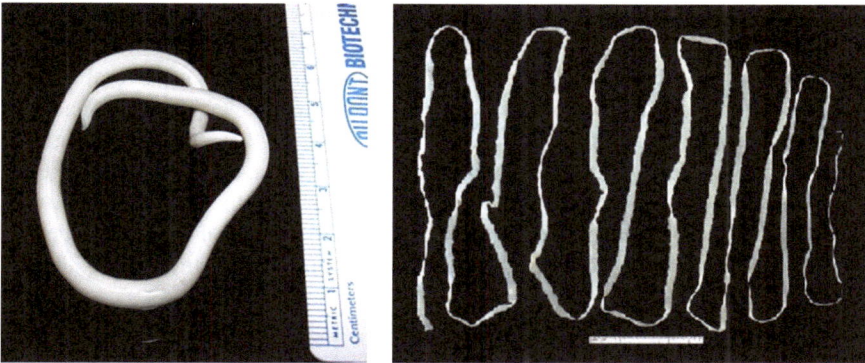

Fig. 3.5 Two common helminths of India: *Ascaris lumbricoides* (*left*) and *Taenia saginata* (*right*) (From the Centers for Disease Control and Prevention)

profiles, potentially helping to improve lipid metabolism and avoid diabetic complications. HECN treatment helped restore elevated levels of particular bio-marker enzymes indicative of impaired liver function—such as serum glutamic oxaloacetic transaminase (SGOT), serum glutamic pyruvic transaminase (SGPT), and serum alkaline phosphatase (SALP)—to normal levels, indicating decreased diabetic complications. Treatment with HECN also inhibited hepatic lipid per-oxidation levels and boosted liver antioxidant parameters towards non-diabetic levels, thereby resulting in less tissue injury and fewer pathologic complications of diabetes.[59] Thus, hydro-methanol extract of *Cocos nucifera* has demonstrated significant anti-diabetic and lipid profile improvement abilities that can be further explored for therapeutic application in humans.

Helminth Treatment

Billions of humans and animals suffer from infections of helminths,[60] which are multicellular parasitic worms that derive nourishment from their hosts.[61] The situation is particularly unfortunate in India, where hundreds of millions of people are estimated to have intestinal nematode infections (nematodes are one type of helminth).[62] Two types of helminths commonly found in India are pictured in Fig. 3.5. This massive burden of helminth infection, particularly on the developing world, has spurred investigations of various ways to prevent and get rid of the parasites.

---

[59] Naskar (2011).

[60] Mehlhorn (2008).

[61] Jaykus et al. (2009).

[62] de Silva et al. (2003).

Research has demonstrated that coconut possesses anti-helminthic properties. One study selected naturally infected sheep in three different locations—Egypt, Germany, and Saudi Arabia—to test the anti-helminthic effects of several crops in differing geographies. After being fed 60 g coconut endosperm powder and 60 g onion powder with their food each day for eight days, the animals with gastro-intestinal nematodes and cestodes no longer had any worm stages in their feces, as observed on days 9 and 20. Compared to untreated sheep, their body weights also increased substantially, indicating that internal worm reduction was very effective. Hence, coconut has promising anti-helminthic applications that could also be explored in humans.[63]

Ulcer Treatment

Coconut milk has been shown to be an effective anti-ulcerogenic agent. Subcutaneous injections of indomethacin led to ulceration in male rats, and the cytoprotective abilities of coconut water, coconut milk, and sucralfate (a conventional cytoprotective medicine used for treating ulcers) were compared. While coconut water resulted in a 39 % reduction in mean ulcer area, coconut milk resulted in a 54 % reduction, which is nearly the same as the 56 % result produced by sucralfate. The difference between the sucralfate and coconut milk results were not statistically significant, so they are considered to have essentially the same effect. Thus, utilizing properties of coconut milk could prove to be an effective means to treat gastritis and ulceration, and its low cost could be more suited to developing countries like India.[64]

Intravenous Treatment

Coconut water is an excellent option for a short-term intravenous hydration fluid, as demonstrated by studies throughout the world. This is particularly true in many areas of the developing world, where more advanced treatment fluids might be too expensive or difficult to access.[65] Found as a free fluid inside the coconut, coconut water is an acidic solution comprised mainly of amino acids, electrolytes, and sugars. Its electrolytic components—rich in calcium, chloride, magnesium, and potassium—are similar to those of intracellular fluid, while its sodium content is

---

[63] Mehlhorn et al. (2010).

[64] Nneli and Woyike (2008).

[65] Campbell-Falck (2000).

lower than human plasma.[66] Such properties indicate coconut water's potential application for total parenteral nutrition (TPN) under particular circumstances.[67]

Coconut water has even been used in large volumes as a resuscitating and hydrating intravenous fluid during conflicts ranging from World War II to the Nigerian civil war of the 1960s.[68] Studies have demonstrated its successful use in cases where infusion over the course of 6 to 12 h totaled 3 liters of coconut water.[69] Since green coconut water is usually sterile, it is well-suited to intravenous use, but it can also be used as a culture medium in microbiology and plant biology.[70]

In humans, the coconut water displayed similarities to a hypotonic intracellular fluid, and possessed slightly more than one half of the typical human blood plasma's cation or anion content.[71] Such high concentrations were found to be quite safe and did not cause any negative coagulation initiation processes.[72] Although coconut water may not be precisely on par with standard TPN solutions, it is a viable low-tech, low-cost solution that could increase health care access if used as an intravenous fluid.

Jackfruit Applications

Protease Inhibitor

Jackfruit is commonly used in folk medicine and herbal treatments. Diarrhea, dermatosis, anemia, and asthma can all be treated with jackfruit leaves and roots, which can also be used as an expectorant for coughs.[73] Oftentimes, biomaterial and therapeutic applications explore the fruit or leaves of a plant. But since jackfruit is a rubber-producing plant, all parts of the jackfruit tree contain a sticky aqueous emulsion—latex—that also has important medical applications.[74]

From the milky latex of the *Artocarpus heterophyllus* fruit stem, researchers have used heat precipitation and anion exchange chromatography to purify a heat-stable heteromultimeric glycoprotein (HSGPL1) that can act as a protease inhibitor. This protein can alter the human blood coagulation system's intrinsic pathways by increasing the activated partial thrombin time (APTT) and inhibiting

---

[66] Pummer et al. (2001).

[67] Petroianu et al. (2004).

[68] Ibid.

[69] Pummer et al. (2001).

[70] Petroianu et al. (2004).

[71] Eiseman (1954).

[72] Pummer et al. (2001).

[73] Fernando et al. (1991).

[74] Mekkriengkrai et al. (2004).

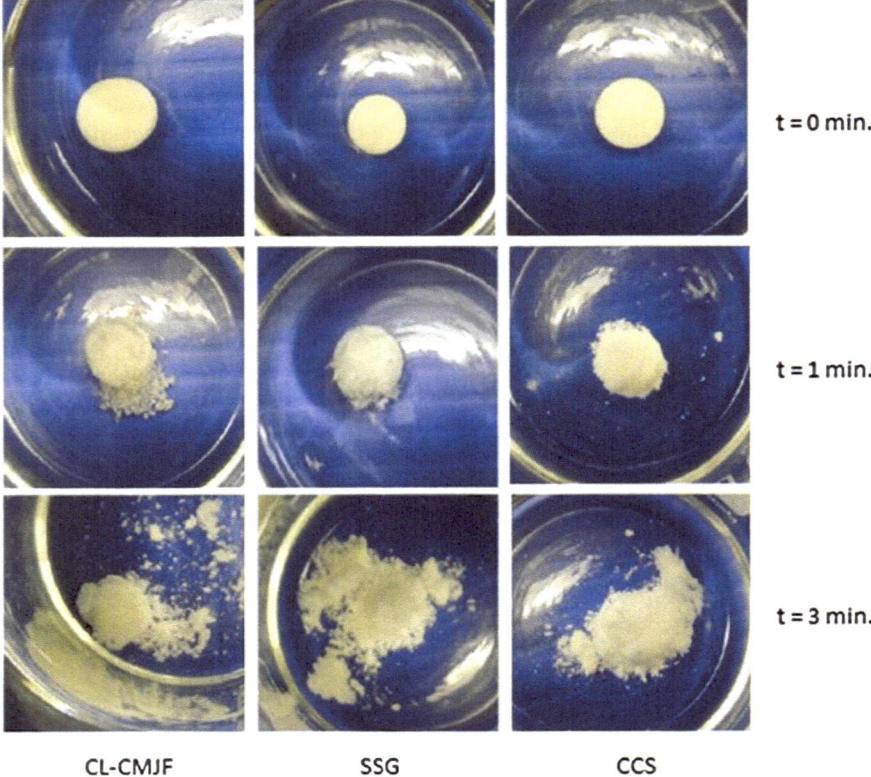

t = 0 min.

t = 1 min.

t = 3 min.

CL-CMJF   SSG   CCS

Fig. 3.6 Comparison of tablet disintegration with disintegrants at 2 %w/w. Kittipongpatana et al. (2011)

particular blood coagulation factors, called XIa and α-XIIa.[75] The process by which human blood clots can be explained via three pathways: intrinsic, extrinsic, and common.[76] APTT is a way to measure the functions of the latter two pathways, representing how long human plasma takes to clot after an intrinsic pathway activator, phospholipid, and calcium have been added.[77] In addition, HSGPL1 contributes to the maintenance of homeostasis in cells under both optimal and adverse growth conditions. These glycoproteins can help proteins refold under conditions of stress and they can also assist with stabilizing proteins and membranes.[78]

---

[75] Siritapetawee and Thammasirirak (2011).

[76] Brown (1988).

[77] Shih et al. (2010).

[78] Siritapetawee and Thammasirirak (2011).

Therefore, this jackfruit-derived protein could be extremely useful for the diagnosis or treatment of hemorrhagic or thrombotic conditions. Potential applications for HSGPL1 include treatment for fibrinolysis (which prevents blood clot formation), wound healing, and blood coagulation.[79] This is especially important given the extremely high cost of many treatments currently offered in the market.[80]

Tablet Disintegrant

Starches derived from jackfruit seeds have potential applications as tablet disintegrants. *Artocarpus heterophyllus* seeds—which are often considered biowaste material—are approximately one-fifth starch on a dry weight basis. This starch possesses characteristics different from many other starches, particularly with respect to acid resistance, granule shape, and thermal and mechanical properties.[81]

In particular, scientists have found that carboxymethl jackfruit starch crosslinked with sodium trimetaphosphate (CL-CMJF) possesses some properties suitable for a tablet disintegrant. Sodium starch glycolate (SSG) and crosscarmellose sodium (CCS) are two common commercial superdisintegrants used today. The CL-CMJF—which can be produced via a dual, simultaneous reaction involving non-toxic chemicals and solvents—proved to be insoluble in water but demonstrated a swelling ability and uptake profile similar to those of CCS.[82] Figure 3.6 shows the disintegration patterns of three tablets containing CL-CMJF, SSG, and CCS over a three minute interval.

Thus, it is worth further exploring the applications of jackfruit seed starch for tablet disintegrants, as they may prove to be a more effective and less expensive option than those currently used.

Leukemia Treatment

Leukemia is the most common form of childhood cancer in India, accounting for between 25 and 40 % of all such cases. With such a low prognosis that only one-third of leukemia-stricken children in India survive at 5 years, leukemia is also the country's largest contributor to cancer-related mortality.[83] Thus there is a pressing need to improve treatment for this cancer to lower related morbidity and mortality.

---

[79] Ibid.

[80] Davis et al. (2011).

[81] Tongdang (2008).

[82] Kittipongpatana et al. (2011).

[83] Arora et al. (2009).

ArtinM is a lectin derived from jackfruit that binds D-mannose and demonstrates high specificity for a trimannoside present in the core of N-glycans located on cell surfaces.[84] Studies have shown that this lectin possesses many properties beneficial to health, such as the ability to speed up the healing of wounds and regeneration of epithelial tissue. It also acts on macrophages and dendritic cells—to establish in vivo Th1 immunity and confer protection against particular intracellular pathogens—as well as on neutrophils, to induce phagocytic and cell-killing activities.[85] Such characteristics led to the trial of ArtinM as a potential anticancer agent against leukemia.

In one experiment, ArtinM was purified by affinity-chromatography from jackfruit seeds and tested against cells from a cell line called NB4 that serves as an appropriate model of acute promyelocytic leukemia (APL). Flow cytometry revealed that ArtinM was able to bind to over 95 % of these leukemia cells, likely due to the modification of NB4's surface glycans caused by malignant transformation. In addition, MTT assays indicated that NB4 cells were quite sensitive to ArtinM inhibition, with a 10 μg/mL concentration of the lectin reducing leukemia cell growth by half. NB4 cells cultured with ArtinM at this concentration for two days, at which point the cells were analyzed by flow cytometry. The lectin caused the pronounced surface exposition of phosphatidylserine, a phospholipid component generally located on the inside of the cell membrane. Hence, this analysis revealed signs of cellular apoptosis. ArtinM treatment also disrupted the mitochondrial transmembrane electrical potential in these leukemia cells.[86]

ArtinM was also compared to all-trans-retinoic acid (ATRA), the main pharmacological component of APL therapy.[87] Unlike drugs like ATRA, ArtinM was able to induce NB4 cell death without cell maturation. In addition, ArtinM treatment generated high levels of reactive oxygen species (ROS), which can lead to depolarization of the mitochondrial membrane translocation of apoptotic factors from the mitochondria to the nucleus. Therefore, the high ROS levels caused by ArtinM treatment can also induce leukemic cell death.[88]

Therefore the jackfruit-derived lectin ArtinM has demonstrated the ability to mediate the death of human leukemia cells, a property that should be further explored in future studies as a potential mechanism by which to treat leukemia.

---

[84] Misquith et al. (1994).

[85] Carvalho et al. (2011).

[86] Ibid.

[87] Segalla et al. (2001).

[88] Ibid.

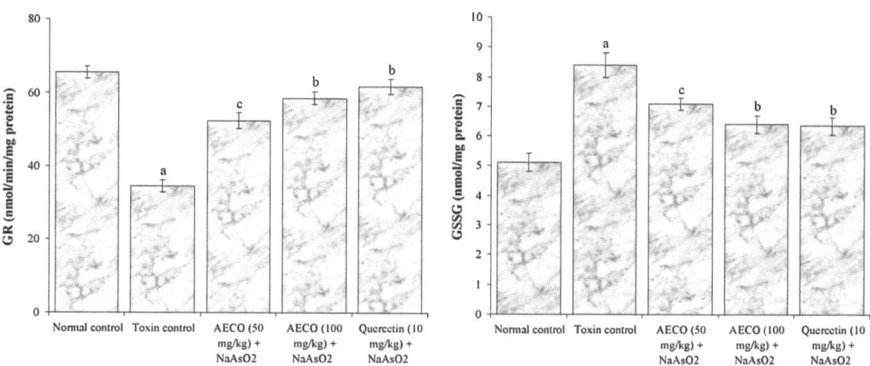

Fig. 3.7 Effect of AECO on GR and GSSG in brain tissue of experimental rats (Das et al. 2009)

Jute Applications

Wound Dressing

Jute has been used as a wound dressing since at least the 1800s, when the textile was used for medical applications much more frequently than today. Nevertheless, jute has reemerged as an effective wound dressing, in the form of a composite with polyvinyl alcohol (PVA). This hydrogel-based option is being developed in Bangladesh as a much more cost-effective option than many synthetic dressings currently available. The final product is intended to be even less expensive than similar hydrogel dressings based on a combination of agar, polyethylene glycol (PEG), and polyvinylpyrrolidone (PVP).[89] In addition, other jute-based dressing options are being explored due to its mechanical properties. Its tenacity, defined as the force required to break the fiber, can be more than three times higher than cotton's and six times greater than that of wool.[90] Such mechanical properties make jute a fiber worth exploring for future wound dressing applications.

Arsenic Protection

Considered "one of the world's biggest natural groundwater calamities," groundwater arsenic contamination in India's Ganga–Brahmaputra plains and Bangladesh's Padma-Meghna plains has been a massive problem ever since the 1983 discovery of arsenic contamination in the region.[91] By the year 2000,

---

[89] IAEA RCA (2003).

[90] Hamlyn and Schmidt (1994).

[91] Ghosh and Singh (2009).

Fig. 3.8 Diagrams
indicating direction of
nutrient and bacterial
transport. Takahashi (2011)

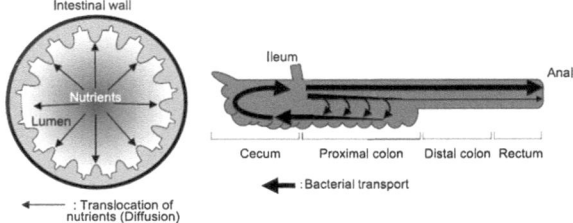

approximately 60 million people in Bangladesh alone were drinking water contaminated with levels of arsenic above the United States EPA limit.[92] Numerous harmful side effects—including cancers, circulatory system problems, nervous system damage, and skin problems—can result from consumption of unsafe levels of arsenic.[93]

One study examined the effects of aqueous extract of *Corchorus olitorius* leaves (AECO) on arsenic-induced brain damage in rats. Five groups of six rats were given daily administrations, respectively, of: double distilled water for 15 days; arsenic (10 mg/kg) for 10 days; AECO (50 mg/kg) for 15 days followed by arsenic (10 mg/kg) for 10 days; AECO (100 mg/kg) for 15 days followed by arsenic (10 mg/kg) for 10 days; and quercetin (a weakly acidic flavonoid commonly found in nature[94]) at 10 mg/kg for 15 days followed by arsenic (10 mg/kg) for 10 days. Their excised brains were then compared.[95]

AECO treatment significantly decreased lipid peroxidation, likely through chemical mechanisms such as free radical quenching, electron transfer, or radical recombination. Probably owing to intrinsic antioxidant properties, AECO also helped combat the changes in levels of brain antioxidant enzymes—including superoxide dismutase (SOD), catalase (CAT), glutathione-S-transferase (GST), glutathione peroxidase (GPx), and glutathione reductase (GR)—that arsenic exposure caused. These antioxidant enzymes are considered the first line of cellular defense against oxidative damage, so their restoration to normal levels is critical. Arsenic administration also negatively affected the levels of reduced glutathione (GSH) and oxidized glutathione (GSSG), cellular metabolites in the brain tissue that act as an additional line of defense. However, AECO treatment before the arsenic exposure helped rats prevent the toxin-induced alteration and maintain levels closer to normal, a result likely owing to the jute extract's free radical scavenging abilities.

Figure 3.7 demonstrates the effect of AECO on two of the compounds previously discussed glutathione reductase and oxidized glutathione. In both cases, the

---

[92] Sambu and Wilson (2008).

[93] Natural Resources Defense Council (2009).

[94] Encyclopaedia Britannica (2012).

[95] Das et al. (2009).

rats treated with AECO demonstrated substantially less damage than in the toxin control, and they were nearly at the same level as the standard quercetin.

In addition, histological observations of brain tissue exposed to arsenic often revealed nuclear pyknosis, a degenerative state that involves chromatin condensation.

However, AECO administration prior to the arsenic treatment substantially lowered signs of pyknosis, left tissue architecture minimally changed, and yielded result comparable to rats treated with quercetin, the positive control.[96] Therefore, despite the brain's generally poor antioxidant defense, jute has demonstrated strong enough protective activity against arsenic-induced oxidative stress to encourage exploration of clinical applications in humans.

Digesta Viscosity Elevation

It is clear to most people that the contents of digesta—food being digested in the stomach—play a crucial role in human health. However, it is important to note that the viscosity of digesta is important as well. Studies have shown that nutrient diffusion in the laminar flow of the intestinal lumen is negatively dependent on digesta viscosity.[97] Higher digesta viscosity also slows the glucose absorption rate in the intestinal lumen of humans and rats, thus lowering the increase in post-prandial blood glucose.[98] This could provide notable health benefits to India's growing diabetic population.

In the large intestine, increasing the digesta viscosity can decrease the fermentation rate by reducing the rate of encounter between bacteria and substrates. The increased digesta viscosity can also reduce diarrhea by decreasing succinic acid.[99] This is significant given the severity of diarrhea as a major global health problem, particularly for children in the developing world. As of 2009, diarrhea was the second largest killer of children under 5 years of age, claiming approximately 1.5 million lives in that category every year (Fig. 3.8).[100]

By supplying solid particles to the digesta, the addition of water-insoluble fibers like crystalline cellulose increases digesta viscosity and makes it harder for nutrients to move through the intestinal lumen and reach the absorptive surface.[101] Insoluble fibers for such applications can be fabricated from the dried leaves of tossa jute.

---

[96] Das et al. (2009).

[97] Antoon and Kirsch (1982).

[98] Takahashi et al. (2005).

[99] Takahashi (2011).

[100] UNICEF/WHO (2009).

[101] Takahashi and Sakata (2005).

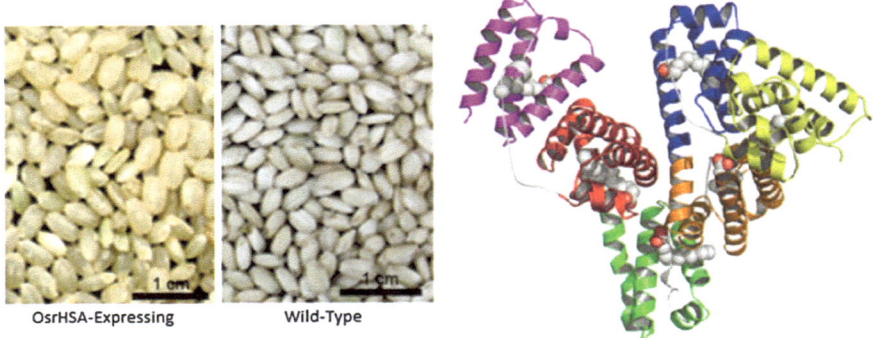

Fig. 3.9 Phenotype of OsrHSA-expressing seeds and wild-type seeds (*left*) and crystal structure of OsrHSA (*right*), He (2011)

When compared to shiitake mushroom (*Lentinus edodes*), a compositionally similar source of insoluble fiber, tossa jute induced a higher coefficient of viscosity at any given shear rate. Tossa jute also demonstrated substantially higher swelling, water-holding, oil-holding, and cation exchange capacities in comparison to shiitake. This is especially noteworthy because insoluble fibers with higher swelling or water-holding capacities have increased digesta viscosities due to reduced digesta free water content. In rats that were administered tossa jute fibers, the gastric, small intestinal, and caecal digesta viscosities were nine, three, and one times (respectively) greater than those of rats that received crystalline cellulose fibers. Hence, insoluble jute fibers have demonstrated their effectiveness in raising digestive viscosity and altering digestion and nutritional absorption as a result.[102]

Rice Applications

Protein Production

Transgenic rice seeds can be used to produce functional human serum albumin (HSA) protein. HSA serves as a monomeric carrier protein for steroids, thyroid hormones, and fatty acids while also helping to stabilize extracellular fluid volume.[103] Its medical applications range from liver cirrhosis treatment to a cell culture medium for vaccine production to a treatment for severe burns.[104, 105] New applications of HSA being explored include carrier of oxygen, fusion of peptides,

---

[102] Takahashi et al. (2009).

[103] Peters (1995).

[104] Alexander et al. (1979).

[105] Hastings and Wolf (1992).

and nanodelivery of drugs.[106] HSA is in high demand worldwide—at over 500 tons per year—but there are several major problems facing its use. Since commercial HSA production is currently based on collected human plasma, the limited supply has led to the production of unsafe, fake albumin.[107] In addition, this method of acquiring plasma-derived HSA (pHSA) carries with it the risk of spreading infectious, blood-derived pathogens such as HIV or hepatitis.[108] Therefore, researchers have been seeking alternate ways to produce safe recombinant HSA (rHSA).

One method to do so involves rice seeds, which are an effective means to produce recombinant proteins due to their ability to generate high levels of stable protein and be controlled well on a production scale. One study used a strong endosperm-specific promoter along with its signal peptide to target rHSA into protein storage vacuoles and enabled transcription of the HSA gene with by using rice-preferred gene codons. This led to *Oryza sativa* recombinant HSA (OsrHSA) expression levels of up to 10.58 % in the endosperm of the transgenic plants. The phenotype and crystal structures of OsrHSA are displayed in Fig. 3.9.

This OsrHSA was found to be equivalent to pHSA with regards to biochemical properties, functions, immunogenicity, and physical structure. With respect to structure and properties, OsrHSA and pHSA had the same amino acid sequence, melting point, molecular mass, and N- and C-terminus. Circular dichroism and spectroscopic analysis also revealed that OsrHSA and pHSA have identical secondary and tertiary structures as well as the same general conformations. In terms of function, OsrHSA and pHSA demonstrated similar ligand binding affinities at site-specific drug markers as well as comparable cell growth promotion across three common cell lines. The two also showed similar efficacy in the treatment of liver cirrhosis, as evaluated in a rat liver ascite model.

Immunoprecipiation experiments and ELISA tests indicate that OsrHSA's immunogenicity in vitro is similar to that of pHSA, and analysis of the antibody titer in rabbit serum after OsrHSA or pHSA immunization indicate that the two also have the same immunogenicity in vivo. The study also demonstrated that producing HSA from transgenic rice seeds was quite inexpensive, particularly compared to current methods. Therefore, OsrHSA is a suitable alternative to pHSA that can be produced from rice seeds on a large scale in a cost-effective manner.[109]

Studies have also shown that it is possible to use rice to produce human lactoferrin (hLF), a multifunctional milk protein involved in biological processes such as bone growth promotion, immune system modulation, iron absorption, and antimicrobial activity. Other transgenic organisms—including cows, fungi, potatoes, and tobacco—have successfully expressed hLF, but rice possesses several

---

[106] Tsuchida et al. (2009).

[107] Xinhua News Agency (2007).

[108] Erstad (1996).

[109] He (2011).

advantages over these other vehicles.[110] The rice system is well-suited to low-cost mass production of hLF and yields expression 25–40 times higher than any other plant system while having the lowest potential (of commonly consumed grains) for causing human allergies.[111] Rice can also stably store expressed foreign proteins in their grains for years. One study showed that 1 hectare of rice could yield up to 40 kg of hLF. In addition, the hLF does not have to be purified from the rice, further simplifying the process to prepare the rice for consumption.

Although this method poses a risk of possible contamination of non-transgenic rice, researchers developed a built-in strategy to combat this threat. They tagged the gene of interest with an RNA interference cassette that would suppress the expression of an enzyme that detoxifies the herbicide bentazon, so hLF could easily be selectively killed using bentazon. Thus, biologically active hLF that is functionally similar to the native human milk protein can also be successfully produced from transgenic rice seeds.[112]

Multifunctional Excipient

Rice germ oil (RGO) is made up of triglycerides and comes from the *Oryza sativa* family.[113] This edible oil is a great source of the antioxidant gamma-oryzanol, which has medical applications ranging from plasma lipid level maintenance to platelet aggregation prevention.[114] Studies have shown that RGO can serve as a suitable multifunctional excipient to carry the active components of a particular medication.

For example, take the case of tacrolimus (TAC), an antibiotic with strong immunosuppressive qualities often administered following an organ transplant to prevent organ rejection. Owing to first-pass metabolism, inter-subject variability, and poor solubility, TAC has very low, variable bioavailability. However, a novel drug delivery system called SMEDDS—self-microemulsifying drug delivery system—has the potential to change this.[115] SMEDDS involves an isotropic combination of drug, oil, surfactant, and co-surfactant that will spontaneously microemulsify upon exposure to gastrointestinal fluid.[116] SMEDDS has some drawbacks, including its restricted solubilization efficiency and its use of antioxidant to inhibit auto-oxidative oil deterioration.[117] Although there are current ways

---

[110] Lin et al. (2010).

[111] Huang (2004).

[112] Lin et al. (2010).

[113] Kim et al. (2002).

[114] Garcia et al. (2009).

[115] Pawar and Vavia (2012).

[116] Constantinides (1995).

[117] Zhang et al. (2004).

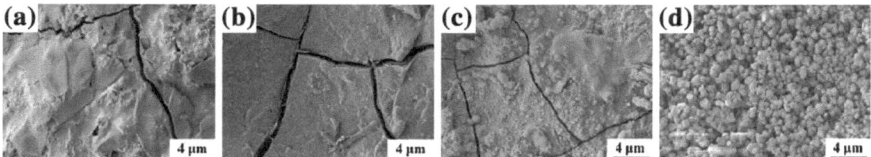

Fig. 3.10 Surface morphology of 75 wt % rice husk scaffolds sintered at 1050 °C for 1 Hr followed by immersion in SBF For: **a** 3 days; **b** 7 days; **c** 14 days; **d** 28 days. Wu et al. (2009)

to combat these limitations, studies have shown that such methods might involve mutagenic and carcinogenic properties.[118] But using multifunctional excipients with high-solubilization efficiency and inherent antioxidant properties could circumvent this problem.

Research has shown that RGO is precisely such a multifunctional excipient, one that can diminish bioavailability and solubility issues that TAC has. In vitro studies showed that a SMEDDS formulation using RGO resulted in the full release of TAC within thirty minutes, while plain TAC and a TAC marketed capsule demonstrated less than 5 and 50 % drug release, respectively, in the same time frame. RGO's natural antioxidant characteristics also prevented auto-oxidative deterioration in TAC-SMEDDS, and the formulation was proven to be stable (over a 3-month period) and compatible with hard gelatin capsules. In addition, TAC-SMEDDS demonstrated a relative bioavailability that was 1.5 times greater than that of the TAC marketed capsule and 3.5 greater than the plain TAC. Maximum concentration displayed a similar effect, with TAC-SMEDDS having concentrations that were 1.69 times higher than the TAC marketed capsule and 8.14 times higher than the plain TAC. Therefore, rice germ oil has demonstrated its promise as a multifunctional excipient for a novel drug delivery system such as that employed in TAC-SMEDDS.[119]

Porous Scaffolds

Good scaffolds for bone tissue engineering should be able to deliver cells while possessing good bioactivity, biodegradability, osteoconductivity, and mechanical properties. They should also have suitable porosity with larger pores to allow tissue ingrowth and new tissue vascularization and smaller pores to encourage protein adhesion as well as cell adhesion and proliferation. Bioactive glasses and glass–ceramics have become an important material for bone regeneration, since they possess excellent bioactivity, controllable biodegradability, osteoconductivity, and the ability to deliver cells. They also have demonstrated the ability to bond

---

[118] Whysner and Williams (1996).

[119] Pawar and Vavia (2012).

to bone in vivo by forming a hydroxyapatite surface layer. However, porosity often poses a challenge.[120]

Rice husk, an abundant material typically regarded as waste, could help solve this problem. In the 2006–2007 year alone, 80 million tons of rice husk were produced, meaning that every 5 tons of rice production yielded approximately 1 ton of rice husk. Rice husk is often used for bedding or burned to generate energy, the latter of which adds to environmental pollution.[121] But scientists have discovered a new biomedical application for this waste material. In bioactive glass and glass–ceramic scaffolds, powdered rice husk can be used as an additive to form pores that will contribute to bone growth.

One study mixed 45S5 Bioglass®—a bioactive glass–ceramic derivative of high-purity SiO_2, Na_2CO_3, $CaCO_3$ and P_2O_5 which is currently used for middle ear and dental implants—with rice husks that had particle sizes below 600 μm to create scaffolds. With blends that contained 70–80 wt % rice husk, scaffolds displayed both macropores (at least 420 μm in length and 100 μm in breadth) and mesopores (between 25 and 80 μm in size).[122] Micropores in the macropore surface increased surface area substantially, thereby facilitating protein adsorption as well as ion-exchange and bone-like apatite surface formation while providing a good microenvironment for cell differentiation and bone matrix deposition.[123]

Figure 3.10 shows the change in surface morphology and increase in surface area after a 75 wt % rice husk blend scaffold is sintered and submerged in simulated body fluid (SBF) for varying amounts of time.

These scaffolds achieved compressive strength in the range of trabecular bone and the scaffolds with higher proportions of rice husk also demonstrated greater mechanical properties.[124] Therefore, rice husk enabled the creation of a porous, bioactive glass–ceramic scaffold for bone regeneration that can provide temporary mechanical support before biodegrading at an appropriate time, and do so in an environmentally friendly and cost-effective manner.

Sutures

Scientists have developed a rice-starch carbon nanocomposite biomaterial that serves as a suturing material. Normally, rice starch can be easily formed into polymer films, but the hydrogen bonds between hydroxyl groups in the matrix of

---

[120] Wu et al. (2009).

[121] Luduena et al. (2011).

[122] Wu et al. (2009).

[123] Kawai et al. (1997).

[124] Wu et al. (2009).

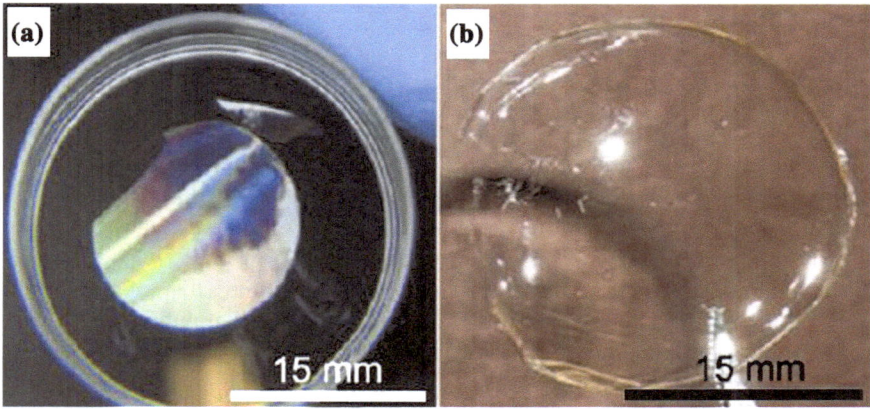

Fig. 3.11 An aligned, grooved, patterned silk film (**a**) and a flat silk film (**b**) (Lawrence et al. 2009)

these films are quite weak. As a result, these natural polymer-based films generally have poor mechanical qualities, particularly with respect to high percentage elongation, tensile strength, and flexural strength.[125] Therefore, it is useful to reinforce rice starch-based films and filaments with other materials to form a more stable composite.

In one study, researchers mixed rice starch in solution with gelatin, sodium carboxymethylcellulose, polyvinyl alcohol, and glycerol before adding carbon powder—derived from coconut shell charcoal –to create a homogeneous mixture. This solution was then dried and formed into stable sutures.[126]

These sutures possess a high water resistance and high mechanical strength while also being biocompatible and bioabsorbable, which makes them good candidates for human body suturing applications. The tremendous improvement in mechanical properties was demonstrated by the fact that a 10 wt % addition of carbon nanopowders to the rice starch polymer film increased the film's elastic modulus by approximately 140 % and raised the tensile strength by about 1200 %. In addition, these rice starch composite sutures are low-cost, environmentally friendly, and relatively simple to manufacture.[127] Thus, they offer many benefits and warrant further exploration, particularly for use in India.

---

[125] Li et al. (2008).

[126] Punyanitya (2010).

[127] Ibid.

Silk Applications

Corneal Grafts

Microbial keratitis is an infection that can progressively damage the corneal epithelium and stroma to the point of scarring, perforation, or blindness. In fact, corneal diseases such as infective keratitis are the second largest cause of vision loss and blindness. This is a particularly large problem in India, where the incidence of corneal infections has been reported at 10 times that of the United States.[128] Corneal blindness is a substantial problem in Indian public health, with many people sustaining corneal injuries from infections, childhood fevers, or chemical injuries that lead to vision loss.[129] Such damage often requires a corneal graft to restore vision. As of 2005, corneal grafts were the most common type of tissue transplant in the US, where nearly 10 million people were suffering from vision loss as a result of corneal blindness.[130] However, within 5 years of implantation, about 1 in 4 patient immune systems rejects the corneal graft.[131]

Silk proteins have significant potential for such corneal tissue engineering applications. In order to develop a new biomaterial option with lower rejection rates, scientists have begun to study silk fibroin, a structural protein from the *Bombyx mori* silkworm cocoon. This protein exhibits controllable degradation rates, non-immunogenic responses in implantation, and robust mechanical properties, making it a suitable candidate for corneal tissue replacement.[132] Silk fibroin is also one of the strongest, toughest natural fibers known to man.[133] Fibroin can be used to create transparent, patterned biomaterials that allow sight and provide a surface to help direct cellular function and matrix deposition.[134, 135] Unlike materials such as synthetic polyglycolic acid or natural collagen, the biodegradation rates of silk fibroin can be manipulated to allow extra time for native tissue to remodel.[136] In addition, biopolymers like collagen and fibrin involve complex processing steps and are difficult to transform into mechanically stable structures, whereas silk is much easier to manipulate and it forms a very robust structure.

The ability to create patterned cell-guiding surfaces as well as porous structures to assist with nutrient diffusion and inter-layer cell interaction further increases the attractiveness of silk-based biomaterials for corneal tissue engineering. The

---

[128] Bharathi et. al. (2009).

[129] Dandona et al. (1998).

[130] Eye Bank Association of America (2005).

[131] Georgea and Larkin (2004).

[132] Lawrence et al. (2008).

[133] Altman et al. (2003).

[134] Jin et al. (2005).

[135] Chirila et al. (2008).

[136] Hu et al. (2005).

patterned films—such as the one pictured in Fig. 3.11—enabled enhanced cell alignment and synthesis of the extracellular matrix, both of which help the biomaterial mimic the native cornea's layered and aligned structures.

In particular, a rabbit corneal cell line expressing green fluorescent protein (GFP-rCF) and human corneal fibroblast (hCF) cell line successfully adhered to, proliferated through, and produced native matrix upon the optically transparent, self-standing silk film substrates. The patterned film helped direct cell and actin filament alignment, and the hCF remained viable and fully proliferated the silk construct within ten days. The silk film structures also exhibited potential to be used as scaffolding in tissue engineering applications, due to the initial native matrix that was produced. Thus, studies have shown the viability of 3D fibroin-based structures to be used as corneal grafts and demonstrated the ability for such silk film constructs to serve as scaffolding for other tissue engineering purposes.[137]

Optical Devices

Today, there are no available optical devices that are mechanically robust while also being fully biodegradable and biocompatible, so many potential medical applications are restricted to retrievable devices. However, silk could change this. Scientists have successfully constructed silk biomaterials with high optical clarity and readout capacity that can achieve complex diffractive structures including lenses and predefined one-dimensional and two-dimensional light patterns. They have created free-standing films between 10 and 100 μm in thickness with transmission across the visible spectrum consistently registering above 90 %. Measured diffraction efficiencies of the silk structure gratings compared favorably to the transmissive glass gratings often used today.[138]

In addition, the unique processing technique required to create these optical silk structures means that they can be designed to more easily incorporate biologically active elements in a manner that poses less of a risk (in terms of potential toxic chemical leaching) to biomedical applications than current systems, which often involve methanol. Research showed that a number of biological substances—including physiologically relevant proteins, enzymes, and small organic pH indicators—could be successfully embedded in the silk films. The introduction of such dopants did not appear to have a significant effect on the film structure, with thickness and refractive indices remaining quite constant.

To test for substrate biocompatibility, red blood cells were included in the silk grating. Hemoglobin, the oxygen-carrying protein inside the red blood cells, was able to maintain activity within the silk structure's matrix, as demonstrated by optical transmission experiments. To study the inclusion of enzymes in the silk

---

[137] Lawrence et al. (2008).

[138] Lawrence et al. (2009).

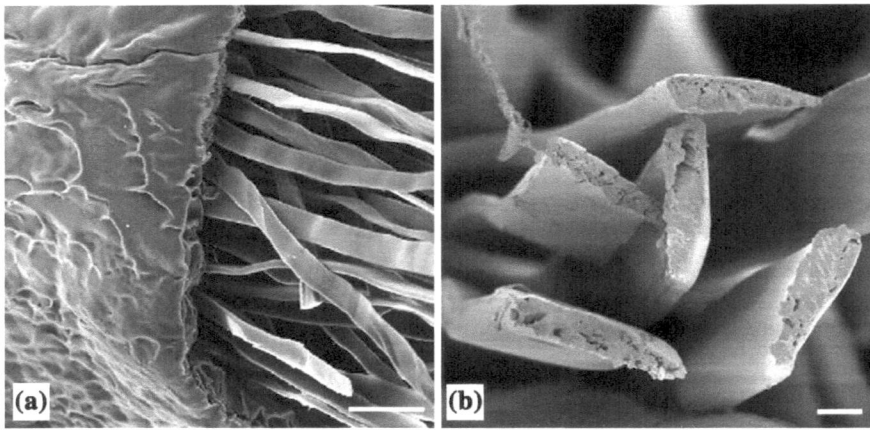

Fig. 3.12 Scanning electron microscope images of a PN200 Spidrex® Silk conduit. Honeycomb-like conduit outer sheath with luminal silk fibers (*scale bar* = 100 µm). Transverse faces of luminal silk fibers at higher magnification (*scale bar* = 10 µm), (Huang et al. 2012)

films, horseradish peroxidase (HRP) was added to the silk solution along with an organic monomer called TMB that changes color in the presence of HRP. Measurements showed that HRP was still active within the silk structure's matrix up to 45 days after initial preparation of the grating. In order to evaluate the silk structure's compatibility with small molecules, Phenol Red, an organic pH indicator, was incorporated. The resulting diffractive optic structures were able to retain both the indicator's functionality and the optical function of the diffraction grating.[139]

The silk structure's biocompatible nature and ability to completely biodegrade is also critical. These characteristics enable the structure's use in devices that could unobtrusively enter and monitor an environment such as a human body. The silk's degradation time could be manipulated such that the device could be used for remote sensing and detection systems but would safely degrade in the environment at an appropriate time, thereby sparing a patient an invasive surgery or other extraction process. Such devices could include everything from an internal glucose meter to hydrogen peroxide detectors. Therefore, silk-based optical devices hold a tremendous amount of promise for future biomedical sensor and detection systems that can be both mechanically durable and fully biodegradable.[140]

---

[139] Lawrence et al. (2009).

[140] Lawrence et al. (2009).

Conduits for Nerve Repair

Another potential application of silk is the creation of artificial conduits for nerve repair. Nearly 3 % of all trauma patients suffer from peripheral nerve injury and regeneration often does not go smoothly.[141] Although the introduction of auto-grafts helped bridge longer-distance nerve grafts without placing the nerve stumps in tension—a problem faced for many years—it still has a range of shortcomings, such as scar tissue invasion, incomplete functional recovery, lack of donor nerves, morbidity from secondary injury, and limited length.[142, 143, 144] Artificial con-duits—which help guide regenerating axons from the proximal to the distal seg-ment—are a promising way to alleviate some of these problems, as they eliminate the risk of secondary injury, reduce scar tissue infiltration, and possess modifiable length, diameter, and permeability. But researchers have been struggling to con-struct the ideal conduit, one that is environmentally compatible and can be easily created in the desired size, that promotes axonal regeneration with a supportive luminal scaffold, and that protects regenerating axons from scar tissue infiltra-tion.[145] The luminal scaffold issue has proven especially tricky, particularly since there are no conduits with luminal scaffolds in clinical use today.[146]

Recently produced silk-based conduits might help solve this problem. These conduits—which consist of a tube of regenerated *Bombyx mori* silk protein with a luminal scaffold of spider silk-like biomaterial filaments—were shown to promote neurite growth.[147] Figure 3.12 shows close up images of such a silk-based conduit. When compared to a non-silk-based nerve guide of poly-3 hydroxybutyrate (PHB) tube, the silk-based conduit performed better in the spinal ganglia of male rats. Approximately 1 month after surgery, axonal labeling mid-conduit was 62 % of the autologous graft, as opposed to 58 % with the PHB conduit.[148] A study using polytetrafluoroethylene (PTFE), silicone, and nerve autograft to bridge a 10 mm gap in a rabbit sciatic nerve showed an average axon count of 48 % in the distal nerve after 13–15 weeks for the PTFE group compared to the nerve graft.[149] Silk conduits, however resulted in 81 % of the number of myelinated axons at 12 weeks, compared to autologous controls.[150]

With respect to biodegradability, three of the five types of conduits presently FDA-approved for clinical nerve repair are either non-biodegradable or degrade

---

[141] Chen et al. (2006).

[142] Battiston et al. (2005).

[143] Chalfoun et al. (2006).

[144] Bellamkonda (2006).

[145] Huang et al. (2012).

[146] Schlosshauer et al. (2006).

[147] Huang et al. (2012).

[148] Hazari et al. (1999).

[149] Azhar and Sara (2004).

[150] Huang et al. (2012).

in as little as 8 weeks, which might be too soon to support axonal growth across long gaps. On the other hand, the silk-based conduits are predicted to be resorbed within 1 year, a timeframe more suited to longer distance repairs. In addition, silk conduits did not trigger a major inflammatory tissue response, which could be a significant advantage over some synthetic nerve guidance channels that elicited a chronic inflammatory response.[151]

Therefore, silk-based conduits could prove to be an excellent candidate for clinical nerve regenerative applications in the future. This is particularly important given that an international panel of nearly fifty experts recently ranked nerve regeneration technologies as the 9th most important regenerative medicine application for improving health in developing countries.[152]

Bone Regeneration

Silk-based scaffolds with a similar lamellar structure as bone have been designed that are useful for tissue engineering and bone regeneration.[153] In particular, clay montmorillonite and sodium silicate on a silk fibroin organic scaffold offers a mechanically stable scaffold for bone growth.[154]

For large bone defects, autologous bone transplants can be used to regenerate the bone, but this method is usually dismissed in favor of biomaterial usage in order to avoid the additional pain, conformal needs at the repair site, longer rehabilitation times, and second site morbidity associated with such transplants.[155] Numerous biomaterials have therefore been studied for this purpose, and collagens have emerged as a suitable biomaterial. Collagens are a natural polymer source that reflect the major source of proteins in the extracellular matrix of bone and interact well with tissue, but they do not possess long-term mechanical stability or integrity.[156] Therefore, silk has emerged as a better possible alternative. Silk possesses extraordinary mechanical properties: for example, silk fibers can absorb approximately the same amount of energy before failure as Kevlar.[157] In addition, purified silk is highly compatible and can demonstrate very slow degradation times.[158] Hence, researchers tested silk fibroin-based scaffolds seeded with human

---

[151] Huang et al. (2012).

[152] Greenwood et al. (2006).

[153] Oliveira et al. (2012).

[154] Mieszawska et al. (2011).

[155] Langer and Vacanti (1993).

[156] Riesle et al. (1998).

[157] Sofia et al. (2001).

[158] Meinel et al. (2005a).

mesenchymal stem cells (hMSC) to promote sustainable bone growth to test their effectiveness in bone regeneration.[159]

These scaffolds created a protein matrix that could slowly degrade and allow for some degree of control over hydroxyapatite bone mineral deposition over time. In bioreactor studies, this enabled the creation of a bone matrix similar to that of spongy bone. According to biochemical assays, gene expression analysis, and X-ray diffractometry, the organic and inorganic components of bone tissues engineered in this manner were quite similar to those of bone. These silk-based, tissue-engineered implants were grown in bioreactors for 5 weeks before introduction into 7-week old mice with circular defects (4 mm in diameter) in their skulls. Within 5 weeks of implantation, the tissue-engineered bone implants demonstrated advanced bone formation. Levels of osteopontin (a sialoprotein that can bind calcium and hydroxyapatite and facilitate cell adherence) and osteocalcin (a secretion of osteoblasts, which are largely responsible for proper bone formation) were also quite high. Thus, in conjunction with engineered bone, silk-based implants have demonstrated sizable promise for effective osteogenesis in a mechanically stable manner.[160]

In addition, sonication-induced silk hydrogels have been created as injectable bone replacements. Studies have evaluated sonication-induced silk hydrogels encapsulating vascular endothelial growth factor ($VEGF_{165}$) and bone morphogenic protein-2 (BMP-2)—two important regulators of angiogenesis and osteogenesis during bone regeneration—for slow release. A study performed on 24 male rabbits with irregularly shaped sinuses showed that these silk-based hydrogels promoted good bone formation and had advantages over current methods, including good plasticity, minimal invasion, simple preparation, and short utility at operation time. In addition, this method was quite simple to use, since its injectable form enabled easier administration given smaller bone windows. The release of $VEGF_{165}$ and BMP-2, which were used to quantify growth factor release from silk gels, demonstrated an impressive ability to resist "bursts" and release the factors according to sustained kinetics over the course of at least 4 weeks. Therefore, silk hydrogels have exhibited potential for use as a minimally invasive means to deliver multiple growth factors via injection to specific locations and assist with bone regeneration in irregular cavities.[161]

Drug Delivery

Beads, hydrogels, and nanoparticles from silk fibroin matrices have also been used for drug delivery.[162] For example, silk fibroin and polyacrylamide semi

---

[159] Meinel et al. (2005b).

[160] Meinel et al. (2005a).

[161] Zhang et al. (2011).

[162] Bhardwaj et al. (2011).

interpenetrating network (semi-IPN) hydrogels have been synthesized to serve as a matrix for sustained drug release with improved mechanical strength and greater drug immobilization capabilities than plain hydrogels. This type of hydrogel was easily fabricated and manipulated with respect to factors such as degradation, mechanical strength, porosity, and swelling. When drug release tested using two model compounds—trypan blue dye and fluorescence isothiocyanate labeled inulin (FITC-Inulin)—the hydrogels demonstrated good release kinetics, though factors such as molecular weight, water solubility, and polymer ratios all influenced the outcomes.[163]

Silk fibroin beads created from scaffolds embedding calcium drug delivery demonstrated the success of silk in dual drug delivery applications. It was able to control and release two model compounds of different molecular weights—bovine serum albumin (BSA) and FITC-Inulin—independently and in a sustained manner. The silk in these systems also helped slow down the drug release to enable sustained delivery and avoid rapid initial bursts. Thus, these silk-based beads have successfully demonstrated a way to deliver multiple active drug molecules from a single delivery system in a regulated manner, a development that could be very important to improvements in modern drug delivery systems.[164]

Spherical silk fibroin nanoparticles for drug delivery purposes have also been formulated from both mulberry silkworms (*Bombyx mori*) and non-mulberry silkworms (*Antheraea mylitta*). Confocal laser scanning microscopy with the FITC-tagged nanoparticles demonstrated their cellular uptake in murine squamous cell carcinoma cells in less than 1 h. Most of these nanoparticles were endocytosed and found in the cytoplasm near the nuclear membrane, where they were able to remain stable. The non-mulberry fibroin nanoparticles were also used to study vascular endothelial growth factor (VEGF) release. VEGF was quite easily loaded into the nanoparticles and the silk allowed for a more controlled, sustained release in comparison to a control system without silk nanoparticles. This demonstrated the nanoparticles' potential for future use in therapeutic drug delivery systems.[165]

Non-mulberry *Antheraea mylitta* silk sericin protein has also been used with amphiphilic poloxamers to create self-assembled micellar nanoparticles for targeted delivery of both hydrophilic and hydrophobic drugs. FITC-inulin was used to study hydrophilic drug carrying abilities. Nanoparticles successfully encapsulated the drug and demonstrated enhanced stability after 10 days of incubation. To observe their behavior as a hydrophobic drug carrier, the nanoparticles were loaded with paclitaxel, an extremely effective anticancer agent. Treatment of breast cancer MCF-7 cells with the paclitaxel-loaded nanoparticles substantially reduced cell growth and increased cell death. Although drug loading was not as high as for free paclitaxel drug delivery, these silk-based nanoparticles may still be preferable because free paclitaxel is less soluble and has a high affinity to other

---

[163] Mandal et al. (2009).

[164] Mandal and Kundu (2009a).

[165] Kundu et al. (2010).

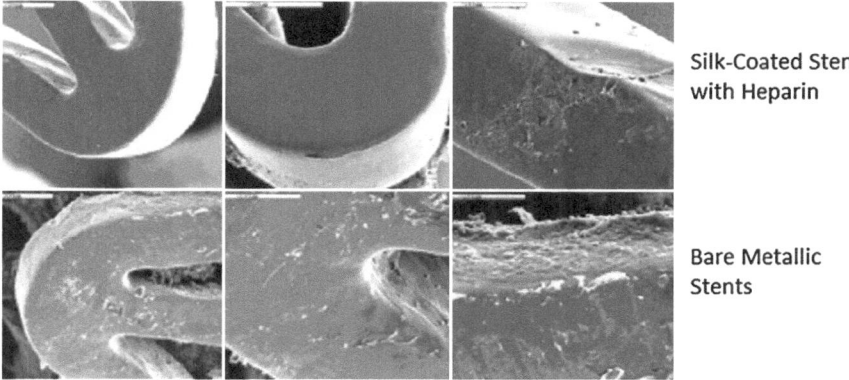

Silk-Coated Stents
with Heparin

Bare Metallic
Stents

Fig. 3.13 SEM micrographs of silk-coated stents with Heparin (*top row*) and bare metallic stents (*bottom row*). Wang et al. (2008)

plasma proteins. Therefore, even at lower concentrations, the drug-loaded sericin nanoparticles can directly enter cells and display functionality, lowering the risk of side effects related to high drug dosages.[166]

Cartilage and Ligament Repair

Given the frequency of injuries and injury-related death due to physical trauma such as traffic accidents, effective ways to generate and repair cartilage in developing countries like India are becoming increasingly important. Pure silk and blended silk fibroin scaffolds have both demonstrated promise for in vitro tissue engineering of cartilage. Several studies have used bovine chondrocytes (the cells that comprise cartilage) to examine properties such as biomechanical strength, cell viability, and proliferation. These studies have revealed the importance of initial cell seeding with mesenchymal stem cells, particularly with respect to density, since the seeding affects extracellular matrix production and biomechanical properties.[167]

Studies have also demonstrated the use of silk scaffolds for ligament tissue engineering. One such study replaced the silk fiber's sericin with gelatin using a cross-linking agent in order to mimic the silk fiber's natural structure and maintain its structural properties like strength while removing the risk of an adverse immune response that sericin could trigger. These scaffolds demonstrated superior

---

[166] Mandal and Kundu (2009b).
[167] IAEA RCA (2003).

mechanical properties and environmental stability when compared with scaffolds made of poly-L-lactide (PLLA) or polylactic-co-glycolic acid (PLGA).[168]

Multi-Functional Stents

Cardiovascular disease has become an increasingly significant problem in India, as it is now the leading cause of death. It is also a substantial source of morbidity, with 9.2 million potentially productive years of life lost in 2000 due to cardiovascular conditions and the number of hypertensive individuals expected to surge to 214 million in 2025.[169] The severity of cardiovascular disease in India has also led to an increase in cardiovascular surgical procedures, including the implanting of stents. In particular, drug-eluting stents have become increasingly common. India has become one of the fastest-growing markets for drug-eluting stents in the world, with over 150,000 such stents implanted in the country every year.[170]

Despite the benefits brought about by stent implantations, the deep vascular injury and endothelial cell damage caused by the procedure in conjunction with long-term exposure to the foreign metallic device often leads to excess vascular smooth muscle cell (SMC) proliferation, thrombosis (formation of a blood clot in a blood vessel or occlusion of the stent),[171] and restenosis (decreased lumen diameter after percutaneous coronary intervention)[172, 173] However, scientists have successfully developed a silk-coated stent capable of releasing drugs over time to alleviate these conditions. Multiple coatings of *Bombyx mori* silk fibroin were loaded with heparin—a naturally occurring anticoagulant also used to coat catheters, metallic stents, and other devices[174]—as well as the chemotherapy drug paclitaxel and the anti-platelet agent clopidogrel to examine their anti-coagulative properties as well as their anti-proliferative effects on smooth muscle cells and endothelial cells. These drug-eluting devices were able to regulate adhesion, viability, and growth of human aortic endothelial cells as well as human coronary artery SMCs in vitro. They also did not compromise on their "primary" function, successfully expanding and holding open the arteries just like regular stents.[175]

In addition, in vivo studies of a porcine aorta revealed that the silk coatings maintained integrity after implantation and were able to successfully reduce

---

168 Liu et al. (2007).
169 Reddy (2007).
170 BioSpectrum Bureau (2011).
171 Mollichelli et al. (2010).
172 Dangas et al. (2010).
173 Wang et al. (2008).
174 Hanson (1998).
175 Wang et al. (2008).

platelet adhesion to the coatings, due to their drug contents.[176] This result is visible in Fig. 3.13, where less platelet adhesion has occurred on the silk-coated stents with heparin.

Hence, these multifunctional stents offer a promising way to efficiently treat cardiovascular disease, and such combination devices could likely be applied to other conditions as well.

Soy Applications

Bone Regeneration

Researchers have created strong, cost-effective composites by combining biodegradable soy-protein polymer and specific bioabsorbable polyphosphate reinforcing fillers.[177] Such biocomposites are nontoxic and bioabsorbable, making them well-suited to medical device applications. These composites are more water-resistant than natural polymers, so they are better able to maintain necessary stiffness in the moist environments present throughout the body.

More specifically, soybean-based biomaterials have demonstrated tremendous potential for bone regeneration purposes. When a bone defect achieves a critical size—as in certain traumatic events or surgical procedures where pathological conditions require bone removal—bone regeneration no longer regenerates spontaneously. Hence there is a need for bone fillers to support bone formation. Human or animal bone grafts (mineralized or non-mineralized) are currently the best option for most of these cases, but such grafts are often in short supply and can increase morbidity and risk of transmittable disease. There are also several categories of synthetic bone fillers which have been shown to support proliferation of bone-producing cells or completely degrade into carbon dioxide and water. However, these two have shortcomings, from brittle mechanical properties to triggering unwanted inflammatory responses. Soybean-based materials offer a simple, inexpensive solution to these problems, as they may exert bioactivity on cells (enhancing tissue regeneration) and they could self-counter any immunogenic response.[178]

One manner in which this has been achieved is by thermosetting defatted soybean curd. The resulting biomaterial is very ductile, enabling effective adaptation to the implantation site and providing the damaged tissue with a biocompatible, biodegradable scaffold. In addition, this soybean-based biomaterial has demonstrated the potential to reduce the chronic inflammatory response induced by macrophages while also promoting bone regeneration by stimulating bone cells. The soybean biomaterial's release of particular isoflavones decreased the activity

---

[176] Ibid.

[177] Otaigbe (1998).

[178] Santin (2007).

of osteoclasts (the cells responsible for bone resorption) and macrophages, thereby speeding up the healing process. Data also indicates that, when combined with induced collagen synthesis, this soybean-based bone filler is able to set off most of the cellular and biochemical events necessary for synthesizing new mineralized tissue.[179] Therefore, this soybean-based biomaterial may prove to be a relatively inexpensive and easily manufactured bone filler with improved biological and physicochemical characteristics for clinical applications.

Other formulations of soybean-based biomaterials have also shown promise as bone fillers.[180] For example, two different soybean-based filler formulations were compared in vivo to a polylactic acid/polyglycolic acid (PLA/PGA) copolymer used in bone grafts as a bone regenerating material for oral surgery or a space filler for tissue regeneration.[181] The first was a 50:50 (by weight) blend of 212–300 μm tofu granules and hydrogel derived from defatted commercial soy flour. The second was a preparation in the proportion of 300 mg soybean-based hydrogel powder with 150 mg soybean granules (of size 212–300 μm) and 100 μm 0.1 M $CaCl_2$ solution. These two soybean-based fillers as well as the commercial PLA/PGA filler were inserted into defects 6 mm in diameter and 10 mm in length in male rabbit femurs.

The soybean-based fillers clearly promoted bone repair and even demonstrated some advantages over the PLA/PGA filler. For example, 8 weeks after implantation, the soybean-based bone fillers promoted a higher level of bone in-growth than the PLA/PGA filler in two-thirds of cases. Also, bone in-growth in the rabbits treated with soybean-based filler progressively improved over the 24 week period of study, while the rabbits treated with the commercial filler displayed clear signs of bone resorption by the eighth week.[182] This resorption is likely due to an inflammatory response triggered by small pieces of degraded PLA/PGA filler, a problem that soybean-based fillers did not encounter.[183] However, the soybean-based biomaterials also encountered some problems. A few rabbits, for instance, demonstrated excessively packed granules. Such models did not permit sufficient bone infiltration and could potentially lead to fibrosis. Nevertheless, this spacing problem is one that could likely be optimized through formula adjustment in order to encourage osteoid infiltration and a progressively mineralizing front. Hence, these soybean-based fillers have demonstrated substantial bone regeneration abilities, likely by using the biomaterial as a scaffold for osteoblasts and inducing cell differentiation. In the future, they could serve as desirable alternatives to methods like autologous bone grafts or PLA/PGA hydrogels, particularly given their ductility and lack of a brittle nature or excessively loose consistency.[184]

---

[179]   Ibid.

[180]   Giavaresi et al. (2009).

[181]   Rimondini et al. (2005), Zaffe et al. (2005).

[182]   Giavaresi et al. (2009).

[183]   Hedberg et al. (2005).

[184]   Giavaresi et al. (2009).

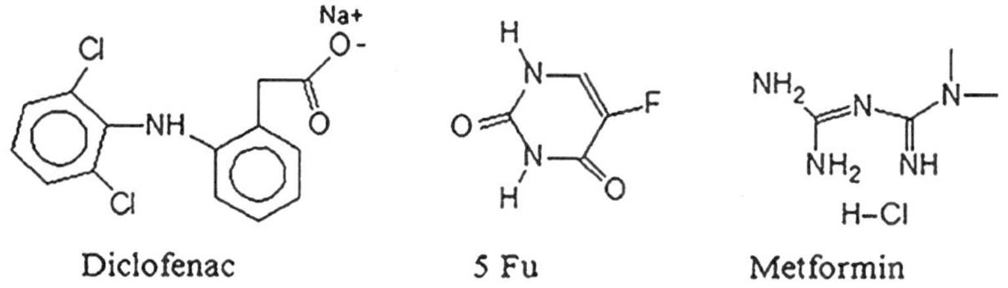

Fig. 3.14 Structures of the three model drugs. Xu and Yang (2009)

Inflammation Treatment

Soy has also demonstrated potential applications in inflammatory disorder treatment. With disorders such as inflammatory bowel disease (IBD), the gastrointestinal tract can be chronically inflamed, leading to such problems as weight loss, gastrointestinal permeability, and digestive tract distress.[185, 186] While there are treatments available for IBD that use drugs and immunosuppressants to target inflammatory molecules, they often have high costs and negative side effects, in addition to occasional low efficacy.[187] Hence, there is currently interest in creating new therapies that are not cytotoxic and can be easily absorbed and disseminated to the site of inflammation while remaining highly effective. Various soy proteins and phytochemicals—including isoflavones, saponins, trypsin inhibitors, and lunasin—have been examined and found to reduce both inflammation and the expression of inflammatory genes in vitro and in vivo.[188]

One study examined the effects of soy-derived di- and tri-peptides on pigs, whose gastrointestinal tracts are morphologically and physiologically similar to those of humans.[189] The pigs were treated with dextran sodium sulfate (DSS) to induce intestinal inflammation and infused with the peptides via catheter. Although treatment with the soy peptides did not substantially affect symptomatic and growth performance factors linked with inflammation, they did help lessen macroscopic indicators of colonic inflammation and intestinal permeability. Peptides exhibited a protective effect in pig colons, helping them maintain an intact epithelium despite the tissue-erosive effect of DSS. Peptide-treated pigs also demonstrated lower colonic crypt depth than those who only received the DSS and similar muscle thickness to control pigs that received neither DSS nor peptides, indicating that the peptides prohibited their intestines from undergoing the typical damage-and-recover cycle of DSS-treated pigs. In addition, soy peptides substantially lowered TNF protein and expression levels. The suppression of TNF has

---

[185] Welcker et al. (2004).

[186] Poritz et al. (2007).

[187] Abraham and Cho (2009).

[188] Young et al. (2011).

[189] Miller and Ullrey (1987).

demonstrated promise for IBD treatment, particularly since intestinal inflammation usually triggers an increase in the expression of TNF and other inflammatory mediators which help release other proinflammatory cytokines, thus drawing out the immune response. The peptides' lowering of levels of the inflammatory mediator IFNG similarly indicates reduced colonic inflammation. Soy-peptide treatment also lowered the levels of cells often associated with Crohn's disease, another common form of IBD.[190]

Thus, soy peptides were able to successfully suppress the expression of inflammatory mediators, as shown by the inhibitory effects on histological measurements and gut permeability as well as the intestines' innate proinflammatory pathways. They have demonstrated sufficient therapeutic potential for patients who suffer from intestinal inflammatory conditions. This is especially significant given the prevalence of ulcerative colitis (one of the most common types of IBD) in India, which is atypically high for the region.[191]

Drug Delivery

Soy beans have high protein content (between 40 and 50 %) and purified soy protein (SP) contains at least 90 % protein on a dry weight basis.[192] This protein can be used to make soy protein films and fibers with useful applications in tissue engineering and drug delivery, a particularly appealing option given their abundant nature and fully biodegradable, biocompatible properties. For example, SP fibers are a useful mechanism for drug sorption and release.

One study prepared SP fibers by dissolving SP in aqueous urea solution with some sodium sulfite, incubating the mixture for two days, and extruding it into a sodium sulfate solution with acetate acid before drawing, drying, and collecting the fibers. Three model drugs—Diclofenac, 5 Fluorouracil (5-Fu), and Metformin—were then loaded onto the SP fibers by both the dissolution and sorption methods. The chemical structures of these drugs are illustrated in Fig. 3.14. The dissolution method involves dissolving the drug in a molten polymer or polymer solution, while the sorption method requires drug sorption into to a fibrous scaffold after its fabrication.[193] The latter method is considered preferable for SP fibers, since it generally leaves behind fewer impurities in the fibers and enables a higher efficiency of drug utilization. Afterwards, drug release was studied in phosphate buffered saline (PBS with pH 7.4) and artificial gastric juice (AGJ with pH 1.2) solutions in a shaking water bath.[194]

---

[190] Young et al. (2011).

[191] Ouyang et al. (2005).

[192] Gibbs et al. (2004).

[193] Prabaharan et al. (2007).

[194] Xu and Yang (2009).

This study showed that drug sorption onto the SP fibers was quite good and increased at higher temperatures. This is likely due to the fact that at increased temperatures, the drugs moved faster, the boundary layer on the surface of the SP fibers thinned, and there were smaller aggregates of drugs, all of which enabled quicker, easier movement of drugs into the SP fibers. Higher temperatures can also increase drug sorption by breaking strong interactions between SP molecular chains, allowing drug molecules to enter. In addition, the study found that 5-Fu and Metformin had higher drug sorption rates than Diclofenac, likely due to their smaller size. This is because smaller drugs have an easier time maneuvering through the openings in SP fibers, so they also experience faster drug sorption at a given temperature.[195]

When compared to polylactic acid (PLA) fibers and starch acetate (SA) fibers, which also have biomedical applications, SP fibers demonstrated faster rates of Diclofenac sorption due to the larger surface openings their fibers possess as well as stronger forces between the drug and SP fibers. SP fibers are also more hydrophilic than those of PLA or SA, meaning that SP swells and its fibers generate larger openings and wider channels, thereby facilitating faster drug sorption rates.[196]

Drug release in PBS was quick, with higher initial bursts from drugs loaded onto SP fibers, but PLA and SA fibers demonstrated more constant subsequent release. However, Diclofenac drug release became much more constant in the AGJ solution (compared to the PBS). Moisture levels played an important role in shaping the constancy of drug release, so that is one factor that could be manipulated in the future to achieve a desired result. Overall, soy protein fibers showed their abilities for drug sorption and release and potential relevant applications in the biomedical field.[197]

Tamarind Applications

Drug Delivery

Biodegradable polymers containing polysaccharides are a good option for nanoparticle carriers in drug delivery. Tamarind seed polysaccharide (TSP) can be obtained from *Tamarindus indica* plants in India and elsewhere in South and Southeast Asia. The chemical structure of TSP is shown in Fig. 3.15. Currently, TSP is used in the food industry as a thickening, stabilizing, and gelling agent.[198] Modifying TSP to become carboxymethyl tamarind kernel polysaccharide

---

[195] Ibid.

[196] Ibid.

[197] Ibid.

[198] Sahoo et al. (2010a).

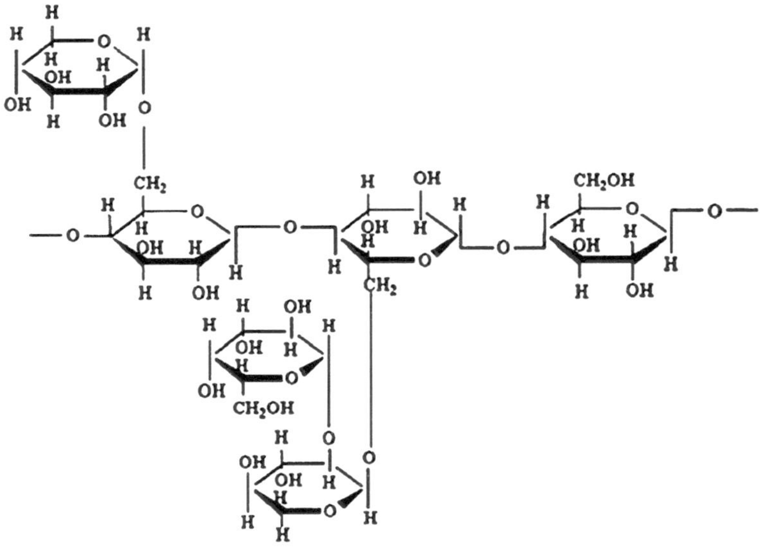

Fig. 3.15 Chemical structure of tamarind seed polysaccharide (TSP). Sahoo et al. (2010a)

(CMTKP) has high potential for pharmaceutical nanotechnology applications due to its greater stability and solubility in cold water. It also possesses a lower biodegradability and longer shelf life.[199] This is particularly important for ocular drug delivery, which is considered one of the most challenging areas to scientists due the eye's "pharmacokinetically critical environment."[200] The eye's numerous barriers—such as the corneal epithelium, iris blood vessels, retinal pigment epithelium, and muco-aqueous layer of the tear film—decrease the amount of each pharmaceutical dose that is actually absorbed.[201] Studies have shown that CMTKP displays excellent ocular tolerability and absorptive properties when tested with tropicamide, an agent often used during eye surgeries and examinations.[202]

TSP has also been studied in the context of delivery of diclofenac sodium, a non-steroidal anti-inflammatory drug often used as a strong analgesic when treating conditions such as rheumatoid arthritis.[203] Although the past few years have seen an increase in attempts to use natural polymers like alginates for drug delivery, such efforts have encountered several problems. For example, with cross-linked alginate hydrogels, the drugs might leak out during gel formation (thereby

---

[199] Kaur et al. (2011).

[200] Duvvuri et al. (2003).

[201] Diebold and Calonge (2010).

[202] Kaur et al. (2011).

[203] Nayak et al. (2010).

decreasing encapsulation efficiency) and the burst of released drugs is quite sharp due to the rapid in vitro degradation release process.[204] However, blends of appropriate polymers—such as those including TSP—can increase drug encapsulation.[205] Researchers have shown that diclofenac sodium-loaded pH-sensitive TSP-alginate composite beads are a suitable method of controlled drug delivery, particularly for pharmaceuticals with various enzymes, peptides, and other physiochemical properties.[206]

Opthalmic Applications

Cataracts, a significant health issue in developed and developing countries alike, are responsible for almost half of all cases of blindness worldwide.[207] The situation is particularly bad in India, where surveys indicate that cataracts account for more than three-quarters of all avoidable blindness.[208] Hence, effective treatments for cataracts are a valuable asset for global health. In addition to intraocular lens replacement, researchers have also been pursuing ways to treat the eye with pharmaceuticals.

The eye is generally impermeable to most environmental agents. The blink reflex assists with a continuous tear flow that clears the ocular surface and inhibits microorganism build-up. With their defensins, lactoferrin, lysozyme, and secretory immunoglobulins, tears also help restrict bacterial growth on the ocular surface.[209, 210] Although most pathogens cannot penetrate a healthy corneal layer, any sort of injury to the ocular surface—caused by anything from allergic hypersensitivity to foreign body trauma—greatly increases the risk of infections like microbial keratitis.[211]

To combat keratitis, high drug concentrations must be maintained at the infected site. This would apply to cataract treatment as well. Ingested drugs make little sense since the cornea is not vascularized, so the most common approach today is to treat keratitis with topical drugs. At the same time, continuous tear flow limits the bioavailability of topical drugs and the corneal epithelium resists drug penetration, so treatment typically involves topical drug administration 1–4 times an hour for 2 or 3 days. Such a treatment generally requires hospitalization, is clearly inconvenient for the patient, and is also associated with toxicity to the

---

[204] Nayak and Pal (2011).

[205] Manjanna et al. (2009).

[206] Nayak and Pal (2011).

[207] Taylor (1999).

[208] Neena et al. (2008).

[209] Haynes et al. (1999).

[210] McClellan (1997).

[211] Sahoo et al. (2011).

corneal epithelium in vitro.[212] However, TSP could change this, given its demonstrated ophthalmic applications.[213]

TSP is branched, non-ionic, and neutral, made up of a cellulose-like backbone with xylose and galactoxylose constituents. This structure lends the polysaccharide a 'mucin-like' molecular configuration, thereby leading to its mucoadhesive properties.[214] TSP's structural similarity to endogenous mucin might be what enables formulations containing the polymer to successfully adhere to the ocular surface for extended periods of time and provide relief for symptoms of dry eye.[215] Some studies have also indicated that TSP might exceed Hyaluronic acid—a visco-enhancer sometimes incorporated in topical eye care solutions—in terms of ocular retention time, relief of dry eye symptoms, and wound healing abilities.[216] In addition, TSP at particular concentrations is able to crystalize in a fern-like shape, increasing its similarity to natural tears.[217]

One ophthalmic application of TSP is as a mucoadhesive polymer that enhances viscosity in order to maximize contact time between antibiotics and the tissue of the cornea. Since TSP is non-toxic to the ocular region, available on the market, a good tear fluid substitute, and an accelerant of the corneal wound healing rate, it is a good option for the ocular administration of antibiotics. Such a use would likely decrease corneal toxicity and increase patient compliance due to its simple and less frequent nature of administration.[218]

The use of TSP for ophthalmic drug delivery is particularly appealing given its ability to release both water-soluble and –insoluble drugs in a controlled manner. The rate of drug release can be easily altered by manipulating the degree of crosslinking or using diluents such as lactose and microcrystalline cellulose.[219]

TSP as a hydrophilic drug delivery system has been proven as an effective model with the use of several drugs, including gentamicin, ofloxacin, nifedipine, and rufloxacin.[220] This demonstrated success as well as TSP's non-carcinogenicity, high drug-holding capacity, and excellent thermal stability all point towards favorable developments of the polysaccharide for biomaterial use in future ophthalmic applications.

---

[212] Sahoo et al. (2011).

[213] Sahoo et al. (2010b).

[214] Sahoo et al. (2011).

[215] Burgalassi et al. (1999).

[216] Mannucci et al. (2000).

[217] Sahoo et al. (2011).

[218] Ibid.

[219] Ibid.

[220] Ibid.

Groups	Conc. (mg/ml)	*Pheretima posthuma*		*Tubifex tubifex*	
		Paralyzing time (min.)	Death time (min.)	Paralyzing time (min.)	Death time (min.)
Distilled water	-	-	-	-	-
Bark extract (alcohol)	5	60.66±0.67*	80.66±0.67*	63.00±2.08*	75.33±1.45*
	10	36.33±0.88*	66.33±0.88*	31.33±0.67*	39.33±0.88*
	15	22.33±0.88^ns	45.00±0.58*	14.66±0.88*	20.66±1.33*
Bark extract (aqeous)	5	>120	NA	67.66±2.60*	81.00±1.00*
	10	89.33±1.76*	115.33±0.88*	53.00±2.52*	74.33±2.08*
	15	58.33±1.45*	87.66±1.45*	23.00±1.73^ns	28.00±1.15*
Leaf extract (alcohol)	5	>120	NA	>120	NA
	10	>120	NA	>120	NA
	15	>120	NA	>120	NA
Leaf extract (aqueous)	5	>120	NA	>120	NA
	10	>120	NA	>120	NA
	15	>120	NA	>120	NA
Piperazine citrate	10	25.00±1.16	64±0.88	22.66±1.76	45.33±1.20

NA = not applicable. Results are expressed as mean ± SEM. N = 3; *P<0.05, as compared to standard, ns = not significant.

Fig. 3.16 Antihelminthic activity of alcohol and aqueous extracts of tamarind leaf, bark. Das et al. (2011)

Anti-Malarial

Malaria is a disease that primarily impacts low-income countries in warm climates that have limited health care facilities. Unfortunately, India fits this description well and the country is estimated to be losing approximately 200,000 lives a year to this illness.[221] Thus it should come as no surprise that health experts are desperate to find ways to suppress this disease and minimize its spread.

In regions as distinct as Ethiopia and India, tamarind has long been used as an antimalarial in folk medicine. Studies have revealed that such traditional treatments are not without base. Water extracts from *Tamarindus indica* fruits demonstrated high chemosuppressive properties, with over four-fifths of the malarial *Plasmodium* present suppressed at a dose of 650 mg/kg.[222] This could be due to the tannins, saponins, sesquiterpenes, alkaloids, and phlobatannins in the tamarind fruit that are thought to have antiplasmodial activity.[223] These studies have shown that tamarind extracts have substantial antimalarial properties that can be further explored and potentially developed into anti-malarial drugs or other therapeutics.

---

[221] Dhingra et al. (2010).

[222] Mesfin et al. (2011).

[223] Doughari (2006).

Anti-Obesity

Tamarind fruits also demonstrate impressive weight-reducing properties. After receiving a high-fat diet to make them obese, rats that were orally administered *Tamarindus indica* pulp aqueous extract (TIE) prepared from matured fruits displayed lower levels of food and caloric intake. Therefore, their mean body weight and body weight gain was far less than in the positive control group of obese rats who did not receive any TIE. In addition to decreasing the weight gain caused by a high-fat diet, TIE has also demonstrated the potential to prevent excessive weight gain.[224] *Tamarindus indica* contains many flavonoid compounds which are considered as inhibitors of fatty acid synthase activity, which could be part of the reason why TIE has displayed such anti-obesity promise. It is also possible that the high acidic and digestible starch contents of tamarind extract alter the colonic mucosa and affect the colon's microfloral content in such a way as to aggravate the adverse effects of a fatty diet and obese condition.[225, 226]

Antihelminthic

The bark of *Tamarindus indica* has demonstrated substantial anthelmintic activity, meaning that it is effective at killing or expelling parasitic worms (helminths).[227] This could be due in part to the tannins present in tamarind bark, which are capable of binding free proteins in the gastrointestinal tract of the host animal or glycoproteins on the parasite's cuticle in order to kill the worm.[228, 229] One study demonstrated the efficacy of tamarind bark extracts in expelling two types of test worms, an Indian earthworm (*Pheretima posthuma*) and an aquarium worm (*Tubifex tubifex*). The earthworm was anatomically and physiologically similar to intestinal roundworm parasites found in humans, and the aquarium worms belong to the same group of Annelida as these earthworms.

As shown in Fig. 3.16, alcohol extract from the tamarind bark at a concentration of 15 mg/mL caused paralysis at 22.33 min and death at 45.00 min for the earthworm, while inducing paralysis at 14.66 min and death at 20.66 min in the aquarium worm. This is very impressive considering each of these times was faster than the times achieved by treatment with 10 mg/mL of piperazine citrate, which was used as a reference standard. (Piperazine citrate causes muscle hyperpolarization that forces relaxation and lowers responsiveness to acetylcholine's

[224] Azman et al. (2011).

[225] Lamien-Meda et al. (2008).

[226] Wang et al. (2006).

[227] Das et al. (2011).

[228] Athanssiadou et al. (2001).

[229] Thompson and Geary (1995).

contractile action, ultimately leading to flaccid paralysis.)[230] Therefore tamarind-based materials have demonstrated sufficient anti-helminthic properties to justify studies for applications in a clinical setting.

References

Abraham C, Cho JH (2009) Inflammatory bowel disease. N Eng J of Med 361:2066–2078

Akao Y et al (2004) A highly bioactive lignophenol derivative from bamboo lignin exhibits a potent activity to suppress apoptosis induced by oxidative stress in human neuroblastoma SH-SY5Y cells. Bioorg Med Chem 12:15

Alexander M et al (1979) Therapeutic use of albumin. J Am Med Assoc 241:2527–2529

Altman GH et al (2003) Silk-based biomaterials. Biomaterials 24:401–416

Antoon C, Kirsch J (1982) Investigation of diffusion-limited rates of chymotrypsin reactions by viscosity variation. Biochemistry 21:1302–1307

Arora R et al (2009) Epidemiology of childhood cancer in India. Ind J Cancer 46:257–259

Athanssiadou S et al (2001) Direct antihelminthic effects of condensed tannings towards different gastrointestinal nematodes of sheep: in vitro and in vivo studies. Vet Parasitol 99:205–219

Azhar M, Sara T (2004) Comparison of nerve graft and artificial conduits for bridging nerve defects. Med J Malays 59:578–584

Azman K et al (2011) Antiobesity effect of *Tamarindus indica* L. pulp aqueous extract in high-fat diet-induced obese rats. J Natl Med 66:22

Bae Y et al (2011) Coconut-derived D-xylose affects postprandial glucose and insulin responses in healthy individuals. Nutr Res Pract 5:5

Battiston B et al (2005) Nerve repair by means of tubulization: literature review and personal clinical experience comparing biological and synthetic conduits for sensory nerve repair. Microsurgery 25:258–267

Bednar H et al (1982) Preliminary findings using wood as an implant material. In: Winter GD, Gibbons DF, Plenk Jr H (eds) Biomaterials 1980. Wiley, New York

Bellamkonda R (2006) Peripheral nerve regeneration: an opinion on channels, scaffolds and anisotropy. Biomaterials 27:3515–3518

Bentley P, Parekh A (1998) Perceptions of anemia and health seeking behavior among women in four Indian states." MotherCare/John Snow, Inc, Washington

Bharathi M et al (2009) Analysis of the risk factors predisposing to fungal, bacterial & Acanthamoeba keratitis in south India. Ind J Med Res 130:749–7

Bhardwaj N et al (2011) Silk protein as biomaterial for tissue engineering and regenerative medicine, New visions for biomaterials and regenerative medicine (Workshop). 16–17 March 2011

Bhaskar J et al (2011) Beneficial effects of banana (Musa sp. var. elakki bale) flower and pseudostem on hyperglycemia and advanced glycation end-products (AGEs) in streptozotocin-induced diabetic rats. J Physiol Biochem 67:8

BioSpectrum Bureau (2011) Boston Scientific launches stent systems in India. BioSpectrum: Asia Edition, 21 February 2011

Brown B (1988) Hematology: principles and procedures. Lea and Febiger, Philadelphia

Burgalassi S et al (1999) Development of a simple dry eye model in the albino rabbit and evaluation of some tear substitutes. Ophthalmic Res 31:229–235

Campbell-Falck D (2000) The intravenous use of coconut water. Am J Emerg Med 18:108–111

---

[230] Das et al. (2011).

Carvalho F et al (2011) The recognition of N-Glycans by the lectin ArtinM mediates cell death of a human myeloid leukemia cell line. PLoS One 6:e27892

Chalfoun C et al (2006) Tissue engineered nerve constructs: where do we stand? J Cell Mol Med 10:309–317

Chandramohan D, Marimuthu K (2011) Characterization of natural fibers and their application in bone grafting substitutes. Acta Bioeng Biomech 13:25

Chandran A et al (2010) The Global burden of unintentional injuries and an agenda for progress. Epidemiol Rev 32:22

Chandy T, Sharma CP (1993) Preparation and performance of chitosan encapsulated activated charcoal (ACCB) adsorbents for small molecules. J Microencapsul 10:475–486

Chandy T, Sharma CP (1998) Activated charcoal microcapsules and their applications. J Biomater Appl 13:128–157

Charnot Y, Peres G (1978) Silicon and bone mineralization. In: Bendx G, Lindquist I (eds) Biochemistry of silicon and related problems. Plenum, New York

Chen M et al (2006) Luminal fillers in nerve conduits for peripheral nerve repair. Ann Plast Surg 57:462–471

Chirila TV et al (2008) Bombyx mori Silk fibroin membranes as potential substrata for epithelial constructs used in the management of ocular surface disorders. Tissue Eng Part A 14:1203–1211

Constantinides PP (1995) Lipid microemulsions for improving drug dissolution and oral absorption: physical and biopharmaceutical aspects. Pharm Res. 12:1561–1572

Dandona L et al (1998) Causes of corneal graft failure in India. Ind J Ophthalmol 46:149–52

Dangas G et al (2010) In-Stent Restenosis in the Drug-Eluting Stent Era. J Am Coll Cardiol 56:1897–1907

Das A et al (2009) Protective effect of *Corchorus olitorius* leaves against arsenic-induced oxidative stress in rat brain. Environ Toxicol Pharmacol 29:6

Das S et al (2011) Determination of Anthelmintic Activity of the Leaf and Bark Extract of *Tamarindus Indica* Linn. Ind J Pharm Sci 73:30

Davis E et al (2011) New anticoagulant therapy for atrial fibrillation and ACS: oral direct thrombin inhibitors. Pharmacotherapy 31:1208–1220

de Silva N et al (2003) Soil-Transmitted Helminthic Infections: Updating the global picture. Disease control priorities Project. July 2003

Dhingra N et al (2010) Adult and child malaria mortality in India: a nationally representative mortality survey. Lancet 376: 1768–1774

Diebold Y, Calonge M (2010) Applications of nanoparticles in ophthalmology. Progr Retin Eye Res 29:596–609

Doughari J (2006) Antimicrobial activity of *Tamarindus indica* Linn. Trop J Pharm Res 5:597–603

Duvvuri S et al (2003) Drug delivery to the retina: challenges and opportunities. Expert Opin Biol Ther 3:45–56

Ebel B et al (2010) Burn injury and the impact on global health. Inj Prev 16:389–392

Eiseman B (1954) Intravenous infusion of coconut water. Am Med Assoc Arch Surg 68:167–178

Encyclopaedia Britannica (2012) Coloration." Encyclopaedia Britannica Online Academic Edition. Accessed 25 March 2012 http://www.brittanica.com/EBchecked/topic/126546/coloration

Erstad B (1996) Viral infectivity of albumin and plasma protein fraction. Pharmacotherapy 16:996–1001

Eye Bank Association of America (2005) Annual report, 2005

Fernando M et al (1991) Effect of *Artocarpus heterophyllus* and *Asteracanthus longifolia* on glucose tolerance in normal human subjects and in maturity-onset diabetic patients. J Ethnopharmacol 31:277–282

Garcia ML et al (2009) Composition, industrial processing and applications of rice bran γ-oryzanol. Food Chem. 115:389–404

Gavin J (2001) Pathophysiologic mechanisms of postprandial hyperglycemia. Am J Cardiol 88:4–8

Georgea A, Larkin D (2004) Corneal transplantation: the forgotten graft. Am J Transpl 4:678–685

Ghosh N, Singh R (2009) Groundwater arsenic contamination in India: vulnerability and scope for remedy. National Institute of Hydrology, Uttarakhand

Giavaresi G et al (2009) Bone regeneration potential of a soybean-based filler: experimental study in a rabbit cancellous bone defects. J Mater Sci 21:22

Gibbs B et al (2004) Production and characterization of bioactive peptides from soy hydrolysate and soy-fermented food. Food Res Int 37:123–131

Gore M, Akolekar D (2003) Evaluation of banana leaf dressing for partial thickness burn wounds. Burns 29:487–492

Gourie-Devi M (2008) Organization of neurology services in India: unmet needs and the way forward. Neurology India 56:4–12

Government of India. "Chapter 7: Nutrition and the Prevalence of Anaemia." From National Family Health Survey-2. Website: http://www.nfhsindia.org/data/india/indch7.pdf

Greenwood H et al (2006) Regenerative medicine and the developing world. PLoS Med 3:e381

Gupta J et al (2010) National programme for prevention of burn injuries. Ind J Plast Surg 43:126–130

Hamlyn P, Schmidt R (1994) Potential therapeutic application of fungal filaments in wound management. Mycologist 8:147–152

Hanson S (1998) Blood-material interactions. In: Black J (ed) Handbook of biomaterial properties. Chapman & Hall, New York

Hastings, G Wolf P (1992) The therapeutic use of albumin. Arch Family Med 1:1023–1026

Haynes RJ, Tighe PJ, Dua HS (1999) Antimicrobial defensin peptides of the human ocular surface. Br J Ophthalmol 83:737–741

Hazari A et al (1999) A resorbable nerve conduit as an alternative to nerve autograft in nerve gap repair. Br J Plast Surg 52:653–657

He Y (2011) Large-scale production of functional human serum albumin from transgenic rice seeds. Proc Natl Acad Sci USA 108:22

Hedberg E et al (2005) In vivo degradation of porous poly(propylene fumarate)/poly(DL-lactic-co-glycolic acid) composite scaffolds. Biomaterials 26:3215

Hsieh M et al (2010) Synthesis, in vitro macrophage response and detoxification of bamboo charcoal beads for purifying blood. J Biomed Mater Res Part A 94A:1133–1140

Hsieh M et al (2007) Application of bamboo charcoal particles in blood purification: cytotoxicity and absorption capability assessments. J Med Biol Eng 27:47–51

Hu X, Lui W, Cui L, Wang M, Cao Y (2005) Tissue engineering of nearly transparent corneal stroma. Tissue Eng 11:1710–1717

Huang N (2004) High-level protein expression system uses self pollinating crops as host. BioProcess Int 2:54–59

Huang W et al (2012) Regenerative potential of silk conduits in repair of peripheral nerve injury in adult rats. Biomaterials 33:59–71

IAEA RCA (2003) Workshop on radiation processing of natural polymers for healthcare applications. http://www.rca.iaea.org/members/Meeting_Conference/RAS/RAS8096/RAS8096%20India%20RWS%20India03.pdf

Ito Y et al (2006) Lig-8, a bioactive lignophenol derivative from bamboo lignin, protects against neuronal damage in vitro and in vivo. J Pharmacol Sci 102:196–204

Jaykus L et al (2009) Food-borne microbes: shaping the host ecosystem. ASM Press, Washington

Jha P et al (2010) HIV mortality and infection in India: estimates from nationally representative mortality survey of 1.1 million homes. Br Med J 340:c621

Jin H et al (2005) Water-stable silk films with reduced beta-sheet content. Adv Funct Mate 15:1241–1247

Joint United Nations Programme On HIV/AIDS (2008) Report on the global AIDS epidemic, 2008

Joshi S, Parikh R (2007) India— diabetes capital of the World: now heading towards hypertension. J Assoc Physicians India 55:323–324

Joshi S et al (2012) Prevalence of diagnosed and undiagnosed diabetes and hypertension in India—results from the screening India's Twin Epidemic (SITE) Study. Diabetes Technol Ther 14:8–15

Joshipura M (2008) Trauma care in India: current scenario. World J Surg 32:1613–1617

Kaiser Family Foundation (2011). The global HIV/AIDS epidemic. Accessed 5 Feb 2012 http://www.kff.org/hivaids/upload/3030-16.pdf

Kaur H et al (2011) Carboxymethyl tamarind kernel polysaccharide nanoparticles for ophthalmic drug delivery. Int J Biol Macromol 50:19

Kawai N et al (1997) Bone formation by cells from femurs cultured among three-dimensionally arranged hydroxyapatite granules. Biomed Mater Res 37:1–8

Kim I et al (2002) Effect of roasting temperature and time on the chemical composition of rice germ oil. JAOCS 79:413–418

Kittipongpatana N et al (2011) Preparation of cross-linked carbomethyl jackfruit starch and evaluation as a tablet disintegrant. Pak J Pharm Sci 24: 415–420

Klemm R et al (2011) Are we making progress on reducing Anemia in Women? The USAID Micronutrient and Child Blindness Project. June 2011

Kosuwon W et al (1994) Charcoal bamboo as a bone substitute: an animal study. J Med Assoc Thail 77:496–500

Kundu J et al (2010) Silk fibroin nanoparticles for cellular uptake and control release. Int J Pharm 388:8

Lamien-Meda A et al (2008) Polyphenol content and antioxidant activity of fourteen wild edible fruits from Burkina Faso. Molecules 13:6

Langer R, Vacanti J (1993) Tissue engineering. Science 260:14

Lawrence B et al (2008) Bioactive silk protein biomaterial systems for optical devices. Biomacromolecules 9:12

Lawrence B et al (2009) Silk film biomaterials for corneal tissue engineering. Biomaterials 30:1299–1308

Lewis D et al (1999) A natural flavonoid present in unripe plantain banana pulp (*Musa sapientum L. var. paradisiaca*) protects the gastric mucosa from aspirin-induced erosions. J Ethnopharmacol 65:7

Li SH et al (1996) Biomimetic coating of bioactives ceramic on bamboo for biomedical applications. J Mater Sci Lett 15:1882–1885

Li SH et al (1997) In vitro calcium phosphate formation on a natural composite material, bamboo. Biomaterials 18:389–395

Li Y et al (2008) Paste viscosity of rice starches of different amylose content and carboxymethylcellulose formed by dry heating and the physical properties of their films. Food Chem 109:1

Lin C et al (2010) A selectively terminable transgenic rice line expressing human lactoferrin. Protein Expr Purif 74:28

Liu H et al (2007) Modification of Sericin-free silk fibers for ligament tissue engineering application. Wiley InterScience

Luduena L et al (2011) Nanocellulose from rice husk following alkaline treatment to remove silica. BioResources 6:1440–1453

Mandal B, Kundu S (2009a) Calcium alginate beads embedded in silk fibroin as 3D dual drug releasing scaffolds. Biomaterials 30:23

Mandal B, Kundu S (2009b) Self-assembled silk sericin/poloxamer nanoparticles as nanocarriers of hydrophobic and hydrophilic drugs for targeted delivery. Nanotechnology 20:11

Mandal B et al (2009) Silk fibroin/polyacrylamide semi-interpenetrating network hydrogels for controlled drug release. Biomaterials 30:8

Manjanna K et al (2009) Diclofenac sòdium microbeads for oral sustained drug delivery. Int J PharmTech Res 1:317–327

Mannucci LL, Fregona I, Di Gennaro A (2000) Use of a new lachrymal substitute (T S Polysaccharide) in Contactology. J Med Contactology Low Vis 1:6–9

Mayo Clinic Staff (2012) "Sepsis." Mayo Clinic Health Information. Accessed 24 March 2012 http://www.mayoclinic.com/health/sepsis/DS01004/DSECTION=causes

McClellan K (1997) Mucosal defense of the outer eye. Surv Ophthalmol 42:233–246

Meagher J et al (2005) Crystal structure of banana lectin reveals a novel second sugar binding site. Glycobiology 15:8

Mehlhorn H (ed) (2008) Encyclopedia of parasitology, 3rd edn. Springer, New York

Mehlhorn H et al (2010) Addition of a combination of onion (*Allium cepa*) and coconut (*Cocos nucifera*) to food of sheep stops gastrointestinal helminthic infections. Parasitol Res 108:1

Meinel L et al (2005a) Silk implants for the healing of critical size bone defects. Bone 37:22

Meinel L et al (2005b) The inflammatory responses to silk films in vitro and in vivo. Biomaterials 26:147–155

Mekkriengkrai D et al (2004) Structural characterization of rubber from jackfruit and euphorbia as a model of natural rubber. Biomacromolecules 5:2013-2019

Mesfin A et al (2011) Ethnobotanical study of antimalarial plants in Shinile District, Somali Region, Ethiopia, and in vivo evaluation of selected ones against *Plasmodium berghei*. J Ethnopharmacol 139:11

Mieszawska AJ et al (2011) Clay enriched silk biomaterials for bone formation. Acta Biomaterialia 7: August 2011 http://www.ncbi.nlm.nih.gov.ezp-prod1.hul.harvard.edu/pubmed/21549864

Miller E, Ullrey D (1987) The pig as a model for human nutrition. Ann Rev Nutr 7:361–382

Misquith S et al (1994) Carbohydrate binding specificity of the B-cell maturation mitogen from Artocarpus Intergrifolia seeds. J Biol Chem 269:30393–30401

Mollichelli N et al (2010) Prolonged Double Antiplatelet Therapy in a Cohort of "De Novo" Diabetic Patients Treated with Drug-Eluting Stent Implantation. Am J Cardiol 105:15

Naskar S (2011) Evaluation of antihyperglycemic activity of *Cocos nucifera* Linn. on streptozotocin induced type 2 diabetic rats. J Ethnopharmacol 138:20

Natural Resources Defense Council (2009) Arsenic in drinking water. Accessed 2 Feb 2009 http://www.nrdc.org/water/drinking/qarsenic.asp

Nayak AK et al (2010) Evaluation of Spinacia oleracea L. leaves mucilage as an innovative suspending agent. Chemistry 19: 338–341

Nayak A, Pal D (2011) Development of pH-sensitive tamarind seed polysaccharide-alginate composite beads for controlled diclofenac sodium delivery using response surface methodology. Int J Biol Macromol 49:23

Nneli R, Woyike O (2008) Antiulcerogenic effects of coconut (*Cocos nucifera*) extract in rats. Phytother Res 22:3

Neena J et al (2008) Rapid assessment of avoidable blindness in India study group. PLoS One 3:e2867

Ojewole J, Adewunmi C (2003) Hypoglycemic effect of methanolic extract of Musa paradisiaca (Musaceae) green fruits in normal and diabetic mice. Methods Find Exp Clin Pharmacol 25:453–456

Oliveira AL et al (2012) Aligned silk-based 3-D architectures for contact guidance in tissue engineering. Acta Biomaterialia 8:1530–1542

Otaigbe J (1998) Controlling the water absorbency of agricultural biopolymers. Plast Eng 54:37

Ouyang Q, Tandon R, Gol KL et al (2005) The emergence of inflammatory bowel disease in the Asian Pacific region. Curr Opin Gastroenterol 21:408–413

Pari L, Umamaheshwari J (2000) Antihyperglycemic activity of *Musa sapientum* flowers: effect on lipid peroxidation in alloxan diabetic rats. Phytother Res 14:1–3

Pawar S, Vavia P (2012) Rice germ oil as multifunctional excipient in preparation of self-microemulsifying drug delivery system (SMEDDS) of tacrolimus. AAPS Pharm Sci Tech 13:10

Peters T (1995) All about albumin: biochemistry, genetics, and medical applications. Academic, San Diego

Petroianu G et al (2004) Green coconut water for intravenous use: trace and minor element content. J Trace Elem Exp Med 17:30

Poritz L et al (2007) Loss of the tight junction protein ZO-1 in dextran sulfate sodium induced colitis. J Surg Res 140:1

Prabaharan M et al (2007) Preparation and characterization of poly(L-lactic acid)-chitosan hybrid scaffolds with drug release capability. J Biomed Mater Res B: Appl Biomater 81 B:427–434

Pummer S et al (2001) Influence of coconut water on hemostasis. Am J Emerg Med 19:287-289

Punyanitya S (2010) Absorbable suture made from rice starch. Adv Mater Res 123–125

Rahman M et al (2006) Acute hemolysis with acute renal failure in a patient with valproic acid poisoning treated with charcoal hemoperfusion. Hemodial Int 10:256–259

Reddy K (2007) India wakes up to the threat of cardiovascular diseases. J Am Coll Cardiol 50:1370–1372

Reed T (2010) Protein in bananas could help block spread of HIV, University of Michigan researchers say. Ann Arbor, 15 March 2010

Riesle J et al (1998) Collagen in tissue-engineered cartilage: types, structure, and crosslinks. J Cell Biochem 71:1

Rimondini L et al (2005) In vivo experimental study on bone regeneration in critical bone defects using an injectable biodegradable PLA/PGA copolymer." Oral Surg Oral Med Oral Pathal Oral Radiol Endotol 99:148–154

Rony K et al (2011) Ganoderma lucidum (Fr.) P. Karst occurring in South India attenuates gastric ulceration in rats. Ind J Nat Prod Resour 2: 19–27

Sahoo R et al (2010a) Release behavior of anticancer drug paclitaxel from tamarind seed polysaccharide galactoxyloglucan. Eur J Sci Res 47:197–206

Sahoo S et al (2010b) Mucoadhesive nanopolymer—a novel drug carrier for topical ocular drug delivery. Eur J Sci Res 46:401–409

Sahoo S et al (2011) Tamarind seed polysaccharide: a versatile biopolymer for mucoadhesive applications. J Pharm Biomed Sci 8:8

Sambu S, Wilson R (2008) Arsenic in food and water—a brief history. Toxicol Ind Heal 24:217–226

Sanghavi P (2009) Fire-related deaths in India in 2001: a retrospective analysis of data. Lancet 373:1282–1288

Sankar M et al (2008) Sepsis in the Newborn. Ind J Pediatr 75(3):261–266

Santin M et al (2007) A new class of bioactive and biodegradable Soybean-based bone fillers. Biomacromolecules 8:7

Schlosshauer B et al (2006) Synthetic nerve guide implants in humans: a comprehensive survey. Neurosurgery 59:740–748

Segalla S et al (2001) Retinoic acid receptor fusion to PML affects its transcriptional and chromatin-remodeling properties. Mol Cell Biol 23:8795–8808

Seri K et al (1996) L-arabinose selectively inhibits intestinal sucrase in an uncompetitive manner and suppresses glycemic response after sucrose ingestion in animals. Metabolism 45:1368–1374

Shih M et al (2010) Characterization of two soy bean (Glycine max L.) LEA IV proteins by circular dichroism and Fourier transform infrared spectrometry. Plant Cell Physiol 51:395–407

Siritapetawee J, Thammasirirak S (2011) Purification and characterization of a heteromultimeric glycoprotein from Artocarpus heterophyllus latex with an inhibitory effect on human blood coagulation. Acta Biochim Pol 58:30

Sofia S et al (2001) Functionalized silk-based biomaterials for bone formation. J Biomed Mater Res 54:3162–3166

Swanson M et al (2010) A lectin isolated from bananas is a potent inhibitor of HIV replication. J Biol Chem 285:19

Takahashi T (2011) Flow behavior of digesta and the absorption of nutrients in the gastrointestine. J Nutr Sci Vitaminol 57:265–273

Takahashi T, Sakata T (2005) Insoluble dietary fibers: the major modulator for the viscosity and flow behavior of digesta. Foods Food Ingred J Jpn 210:944–953

Takahashi T et al (2005) Crystalline cellulose reduces plasma glucose concentrations and stimulates water absorption by increasing the digesta viscosity in rats. J Nutr 135:2405–2410

Takahashi T et al (2009) Water-holding capacity of insoluble fibre decreases free water and elevates digesta viscosity in the rat. J Sci Food Agric 89:177–194

Taylor H (1999) Epidemiology of age-related cataract. Eye 13:445–448

Thompson D, Geary T (1995) The structure and function of helminth surfaces. In: Marr J (ed) Biochemistry and molecular biology of parasites. Academic Press, New York

Tongdang T (2008) Some properties of starch extracted from three Thai aromatic fruit seeds. Starch/Starke 60:199–207

Tsuchida E et al (2009) Artificial oxygen carriers, hemoglobin vesicles and albumin-hemes, based on bioconjugate chemistry. Bioconjugate Chem 20:1419–1440

UNICEF/WHO (2009) Diarrhoea: Why children are still dying and what can be done. http://whqlibdoc.who.int/publications/2009/9789241598415_eng.pdf

United Nations Development Project, India (2011) Responding to the HIV/AIDS challenge. Accessed 23 March 2012

Varki A et al (eds) (1999) Essentials of glycobiology. Cold Spring Harbor Laboratory Press, Cold Spring Harbor

Wang Y et al (2006) Potent inhibition of fatty acid synthase by parasitic loranthus [Taxillus chinensis (dc.) Danser] and its constituent avicularin. J Enzym Inhib Med Chem 21:87–93

Wang X et al (2008) Controlled release from multilayer silk biomaterial coatings to modulate vascular cell responses. Biomaterials 29:894–903

Watts C (2002) The public health impact of microbicides: model projections. http://www.itg.be/micro2002/Pages/Abstracts.html

Welcker K et al (2004) Increased intestinal permeability in patients with inflammatory bowel disease. Eur J Med Res 9:29

Whysner J, Williams GM (1996) Butylated hydroxyanisole mechanistic data and risk assessment: conditional species-specific cytotoxicity, enhanced cell proliferation, and tumor promotion. Pharmacol Ther 71:153–191

World Health Organization (2010) Injuries and violence: the facts. Accessed 22 March 2012 http://whqlibdoc.who.int/publications/2010/9789241599375_eng.pdf

Wu S et al (2009) Preparation of porous 45S5 Bioglass®-derived glass-ceramic scaffolds by using rice husk as a porogen additive. J Mater Sci 20:22

Xinhua News Agency (2007) Drug watchdog tightens supervision of albumin medicine. April 4, 2007. Beijing Review. http://www.bjreview.com.cn/health/txt/2007-04/04/content_60856.htm

Xu W, Yang Y (2009) Drug sorption onto and release from soy protein fibers. J Mater Sci: Mater Med 20:17

Yasuda K (2007) Effects of inulin on iron utilization by young anemic pigs and implications for human nutrition. A thesis presented to the Faculty of the Graduate School of Cornell University in Partial Fulfillment of the Requirements for the Degree of Master of Science. August 2007

Young D et al (2011) Soy-derived di- and tripeptides alleviate colon and ileum inflammation in pigs with dextran sodium sulfate-induced Colitis. J Nutr 142:21

Zhang Q et al (2004) Preparation of nimodipine-loaded microemulsion for intranasal delivery and evaluation on the targeting efficiency to the brain. Int J Pharm 275:85–96

Zhang W et al (2011) The use of injectable sonicatin-induced silk hydrogel for $VEGF_{165}$ and BMP-2 delivery for elevation of the maxillary sinus floor. Biomaterials 32:1

Zaffe D et al (2005) Histological study on sinus lift grafting by Fisiograft and Bio-Oss. J Mater Sci: Mater Med 16:789–793

Part III

Chapter 4
Implications, Challenges, and Recommendations

Summary

Naturally based biomaterials and therapeutics have demonstrated significant promise for biomedical use, particularly in developing countries such as India. Although research has focused primarily on applications for the developed world, biomaterial and therapeutic use could help improve health outcomes in developing regions as well. The development of low-cost bio-derived materials (such as the banana leaf dressing described previously) could help lower the barrier to proper health access and care by making treatment more affordable. Designing biomaterials and therapeutics to be mechanically robust and biocompatible would minimize the need for additional procedures to replace or treat the materials, a process that can be very costly and require time and trained staff beyond the level that is available in many areas of the world. Infection-resistant bio-derived materials that are heat- and time-stable would also be a great step forward, allowing biomaterials and therapeutics to overcome the challenges presented by improper handling due to untrained staff or lack of appropriate facilities. Developing naturally sourced biomaterials and therapeutics that are easy to create or use would also help with these particular issues.

As discussed in the previous chapter, there are clear applications for at least nine common natural resources in India: bamboo, banana, coconut, jackfruit, jute, rice, silk, soy, and tamarind. These resources can be developed into biomaterials and therapeutics with such varied functions as anti-HIV agents, bone growth scaffolds, drug delivery hydrogels, nerve conduits, and tablet disintegrants. Such applications have tremendous potential to help address the current and future health needs of India, which encompass a broad array of communicable and non-communicable conditions. For example, tamarind-based anti-helminthic therapeutics could greatly alleviate the suffering of the hundreds of millions of Indians

V. Eswarappa and S. K. Bhatia, *Naturally Based Biomaterials and Therapeutics,*
SpringerBriefs in Public Health, DOI: 10.1007/978-1-4614-5386-4_4, © The Author(s) 2013

estimated to have intestinal nematode infections.[1] Tamarind saccharide polymer based treatments could help restore vision to those blinded by cataracts, a condition likely to become more prevalent in India given the country's aging population and growing incidence of diabetes.[2]

It is particularly fitting that such biomaterial and therapeutic uses are so well-suited to the country credited with the first written reference of a prosthetic device. The Rig Veda, an ancient Indian poem dating to approximately 3500 BC, includes the story of Queen Vishapala, a warrior queen who loses her leg in battle but is fitted with a prosthetic iron leg and returns to fight.[3] Just as Indians in ancient times were able to develop and use artificial limbs to improve their quality of life, Indians today can harness the power of naturally derived biomaterials and therapeutics to improve health outcomes throughout the population.

Objections and Challenges

Despite the array of potential benefits discussed above, there are several objections to and challenges facing accelerating the use of naturally sourced biomaterials and therapeutics. One argument against the use of such bio-derived materials is that it would not be appropriate given the nutritional state of India today. With over 200 million of its people going hungry in 2011, India ranked an abysmal 98th out of 118 countries on the global hunger index. India also has the highest burden of child malnutrition in the world. To make matters worse, the World Bank warned that three-fifths of the country's food subsidies do not even reach the poor.[4] Would it really be acceptable to use crops for biomaterials and therapeutics when people are starving?

This is a fair argument, especially given that the majority of India's population—as well as the world's—depends upon rice as a staple food.[5]. Therefore, it makes the most sense to concentrate future research on biomaterial and therapeutic applications of resources that are not primarily used for culinary purposes, such as bamboo, jute, and silk. Inedible portions of crops—such as banana pseudostems, coconut shells, and rice husks—should also be a research priority, given that often such materials are otherwise treated as waste.

Another possible objection to using naturally derived biomaterials and therapeutics lies in the fact that many such approaches have not been fully proven yet. Is it worth it to invest time and effort to pursue the development and application of these alternatives when there may be "tried-and-true" options available? For

---

[1] de Silva et al. 2003.

[2] Raman et al. 2010.

[3] Britt, The history of prosthetic devices.

[4] Bhowmick 2011.

[5] International Rice Research Institute 2012.

example, why should one look into bamboo as a bone substitute when there are already bioceramic, metallic, and polymeric biomaterials for the same purposes that work adequately well?

The mere existence of ways to address a problem should not in itself preclude the pursuit of alternate techniques to address the same issue, however. Suppose we applied this argument to other aspects of life beyond the biomedical. Surely fossil fuels are an acceptable way to run our power plants and vehicles, as they help us achieve our ultimate goal of generating electricity or transportation from one location to another. Similarly, oil lamps were generally considered sufficient until the nineteenth century, since they helped illuminate dark areas and some could even be transported where light was necessary. Why, then, did people pursue and largely switch over to electric methods to satisfy their lighting needs? And why are we now seeking to develop alternative energy sources such as wind and solar power to circumvent the use of fossil fuels?

The simple answer is that none of these technologies are perfect. Just as fossil fuels may be problematic due to limited availability or environmental effects and oil lamps may be less than ideal due to fluctuating light intensity or a perpetual need to be refilled, many biomaterials currently in use are problematic due to insufficient biocompatibility, inappropriate mechanical properties, or other factors. Exploring the use of bio-derived materials could help provide solutions to some of these problems. Bamboo charcoal beads, for instance, do not pose the same blood contamination risk during hemoperfusion that commonly used forms of activated carbon do. In addition, developing naturally based biomaterials and therapeutics may lead to additional benefits and applications that were previously unseen. This failure to be complacent and accept current methods as the ideal can help foster an environment of innovation and technological development with positive benefits that extend beyond the reach of bio-derived materials.

An additional challenge is that any naturally derived biomaterials or therapeutics must be proven to be effective and safe before they can be used in the human population. However, just as the U.S. Food and Drug Administration regulates and supervises food, pharmaceuticals, medical devices, and other products to ensure the American public's health and safety, the Central Drugs Standard Control Organization performs a similar function for India. It is important to note that this process can be expensive and time-consuming, particularly given the 10–12 years and $1.5 billion typically required to develop a new drug.[6] Nevertheless, it is vital to ensure the welfare of the Indian population. Seeking approval of developed biomaterials and therapeutics can also help avoid situations of ethically questionable or unsafe testing or administration, such as recent occurrences of inappropriate testing on children, treatment with unapproved methods, or coercion of the poor to be test subjects.[7]

---

[6] Shukla and Sangal 2009.

[7] Glickman et al. 2009.

Another challenge that naturally derived biomaterials and therapeutics face is that they must demonstrate their cost-effectiveness. A bio-derived material that has a slight advantage over an existing device may not necessarily be worth it if its cost is far greater. This is particularly true for a low-income country like India that already has relatively small health expenditures. Therefore, it may be useful to project expenses and conduct cost-benefit analyses in comparison to existing materials to decide which opportunities may be worth pursuing in financial terms. Such studies should be careful to account for other elements beyond the price tag of the final biomaterial or therapeutic. Costs for maintenance, repair, and replacement should be incorporated as well. Cost comparisons should also seek to address variations in performance, safety, comfort, and other properties, despite the greater challenge in assigning a monetary value to such factors.

Finally, it is important to note that even if a country like India successfully derived biomaterials and therapeutics from its own natural resources, it would likely still rely on other countries—particularly those in the developed world—for other materials or components to use. For example, a bio-derived hydrogel might be loaded with a drug produced in the developed world, or a naturally based stent might require tools imported from a developed country to surgically implant the device. Some might argue that this does little to ameliorate the current predicament in which developing nations largely depend on developed countries for assistance with their health problems. Many medical charities based in the western world operate in a "dislocated" manner, giving developing countries equipment or medicines but little else. This only addresses a small fraction of the problematic forces at work in these countries and leaves developing countries with a substantial information and practical knowledge gap.[8] Thus developing nations end up handicapped by their reliance on outside sources for physical goods necessary for health care and rarely learn to become self-sufficient or develop solutions of their own. However, a situation of partial dependence is clearly preferable to one of complete dependence. Such an environment would allow the developing country—in this case, India—to better suit the technologies to their own needs, instead of simply receiving and implementing identical transplanted technologies.

The additional research and infrastructural capacities India would develop as a result of increased bio-derived materials studies and applications would also bring many benefits and likely raise research standards overall. Countries that make serious efforts to build up local research capacity are likely to successfully absorb technologies in related fields at a faster rate,[9] so further exploration of bio-derived materials is likely to speed up the successful implementation of related health technologies. Any benefits to health that bio-derived materials bring to India could also likely be applied to other areas of the world, thereby facilitating a greater exchange of information and ideas to improve health outcomes elsewhere as well.

---

[8] Kaunonen 2010.

[9] Juma 2011.

Moving Forward

Now that we have examined the potential objections and challenges as well as the benefits of pursuing naturally based biomaterials and therapeutics for use in India, we should consider the best path to take moving forward. Although there are many directions to choose from, some are more likely to lead to improved health outcomes in India than others.

The most important next step is to encourage further research, development, and use of naturally derived biomaterials and therapeutics in the developing world as well as the developed world. As shown in the previous chapter, this could help address numerous health concerns in India today and improve upon current methods of treatment and healthcare.

In India, organizations such as the Society for Biomaterials and Artificial Organs could take a leading role in this department. Founded in 1986, this organization has worked to encourage research on biomaterials and artificial organs and bring together professionals from these fields to foster productive work and collaborations in the future. It publishes a biannual journal to showcase relevant work and members regularly contribute to other reputable journals as well.[10] Institutions of higher education could also play a critical role in this pursuit. Indian universities such as the world-renowned Indian Institutes of Technology (IITs) or the Sree Chitra Tirunal Institute for Medical Sciences and Technology have already made substantial contributions to the field of biomaterials and therapeutics. For example, researchers at IIT Bombay have developed novel methods for sterilizing biomaterials using supercritical fluids and designing vesicles via stationary phase inter-diffusion.[11] Meanwhile, faculty members at IIT Delhi have worked on projects ranging from the development of antimicrobial sutures to major organ reconstruction.[12]

A tremendous impact could be created if such societies and institutions were to place a greater emphasis on bio-derived materials. It is likely that other laboratories and organizations would follow suit, creating a fertile environment for the rapid development of effective naturally based biomaterials and therapeutics.

Government action could also spur this increase in bio-derived materials research. For example, it could provide additional research funding to institutions pursuing studies of these areas. This would be similar to the funding provided to United States universities and research institutions by the National Institutes of Health (NIH) for medical research it deems important. Such research could also include cost-effectiveness analyses adhering to the parameters for such studies as detailed previously. In addition, the Government of India could offer subsidies to people and institutions that choose to purchase naturally based biomaterials and therapeutics over materials that are not bio-derived, particularly if the naturally

[10] Society for Biomaterials & Artificial Organs (India) 2006.

[11] Indian Institute of Technology, Bombay 2012.

[12] Indian Institute of Technology, Delhi 2012.

derived options are more expensive. This method of encouraging bio-derived materials use can be compared to the United States Government's subsidies of hybrid vehicle purchases. Both methods of government action could facilitate a rapid increase in the development and application of naturally based biomaterials and therapeutics.

It is also important to explore the possibility of combination devices created from naturally derived materials. Such devices would serve more than one purpose and help facilitate a transition in the perception of biocompatibility from a concept of simply "doing no harm" to one of actively doing good.[13] Although it is great to have biomaterials and therapeutics that are nontoxic and able to maintain the body's current situation without causing additional damage, it would be even better to have materials that would improve upon the body's health status and encourage positive healing responses.

One such example would be combination devices like the drug-eluting stent described in the previous chapter. This silk fibroin-based stent was able to expand normally and successfully support the artery while also releasing drugs to treat restenosis and thrombosis. Such a stent could greatly improve patient health by decreasing side effects and minimizing the inconvenience of taking additional medications while still providing the regular benefits of a normal artery-expanding stent. The development of additional combination devices in the future offers a similar promise of improved means by which to achieve better patient health outcomes.

Finally, it is important that bio-derived materials research further explore "smart" applications, where the biomaterial or therapeutic could sense and react to information being provided in its environment. The silk-based technology for biosensor needs described earlier is a step in the right direction. The development of "smart" biomaterials and therapeutics would enable the provision of more accurate and tailored medical care without the additional human and financial capital such an endeavor usually requires. This would be a particularly useful development for a country like India that has limited resources.

In 1987, a professor named Anthony Gristina predicted that "[u]ltimately, almost every human in technologically advanced societies will host a biomaterial."[14] Although we have not yet reached this point, our world has surely moved closer to such a state, with increasing numbers of people receiving biomaterial implants ranging from pacemakers to bone fillers. Society's use of therapeutics has grown over time as well, particularly with increasing numbers of people taking chemotherapy drugs or medicines to lower high blood pressure or cholesterol levels. In developing and transitional countries, prescriptions of generic drugs increased steadily from the 1980s to 2006.[15]

---

[13] Helmus et al. 2008.

[14] Gristina 1987.

[15] World Health Organization and Harvard Medical School 2009.

These increases in biomaterial and therapeutics use reflect their growing importance in the world today. But the next area for rapid exploration and growth in this field lies in materials that are naturally based. Such progress also requires a shift in geography from the developed world to developing nations like India.

However, the development of bio-derived materials for medical applications is clearly not a panacea for the developing world. No banana leaf dressing or silk-based corneal graft will cause India's plethora of health problems to vanish overnight. Nevertheless, further exploring applications of naturally based biomaterials and therapeutics can help India move towards a more independent, self-sufficient state with respect to biomedical technology. Such a state would not necessitate a return to the pre-economic liberalization days of India before the 1990s. It would simply mean a shift towards a more collaborative environment with respect to health, one that did not rely so heavily on pre-packaged health applications and medical technologies from the western world.

In addition to the sociological benefits and sense of empowerment conferred by a country's ability to help alleviate its own conditions, a 2005 UN report found that it is important to focus on domestic innovation (instead of simply accepting external information and technology imports) since domestic work is more likely to address local health needs and contribute to economic and health development. Other studies have also shown that domestic innovations in the developing world can lead to more affordable treatments.[16] Hence, the further exploration of naturally based biomaterials and therapeutics in India could lead to substantial improvements in health outcomes for much of the population while also leading the country one step closer to a more modern version of Gandhi's *swadeshi*, or self-reliance.

References

Bhowmick N (2011) India's food crisis has many ingredients, The Guardian. 19 July 2011. http://www.guardian.co.uk/global-development/poverty-matters/2011/jul/19/india-food-crisis

Britt, The history of prosthetic devices. The University of North Carolina at Chapel Hill. http://www.unc.edu/~mbritt/Prosthetics%20History%20Webpage%20-%20Phys24.html

Glickman S et al (2009) Ethical and scientific implications of the globalization of clinical research. New Engl J Med 360:19

Greenwood H et al (2006) Regenerative medicine and the developing world. PLoS Medicine 3:1496–1500

Gristina A (1987) Biomaterial-centered infection: microbial adhesion versus tissue integration. Science 237:25

Helmus M et al (2008) Biocompatibility: meeting a key functional requirement of next-generation medical devices. Toxicol Pathol 36:70–80

Indian Institute of Technology, Bombay (2012). Patents classified according to research areas. Department of Chemical Engineering. www.che.iitb.ac.in/online/services/patents/patents-classified-according-research-areas. Accessed 22 Mar 2012

---

[16] Greenwood et al. 2006.

Indian Institute of Technology, Delhi (2012). Faculty, Department of Textile Technology. http://paniit.iitd.ac.in/textile/public/community/faculty/faculty.php. Accessed 22 Mar 2012

International Rice Research Institute (2012). Rice basics. http://irri.org/about-rice/rice-facts/rice-basics. Accessed 15 Feb 2012

Juma C (2011) The new harvest: agricultural innovation in Africa. Oxford University Press, New York

Kaunonen G (2010) Engineering world health. IEEE Pulse, September/October 2010

Raman R et al (2010) Prevalence and risk factors for cataract in diabetes: Sankara Nethralaya Diabetic Retinopathy Epidemiology and Molecular Genetics Study, Report No. 17 Investigative ophthalmology & visual science, vol 51:6253–6261

Shukla N, Sangal T (2009) Generic drug industry in India: the counterfeit spin. J Intellect Prop Rights 14:236–240

de Silva N et al (2003) Soil-transmitted helminthic infections: updating the global picture. Disease Control Priorities Project, London

Society for Biomaterials & Artificial Organs (India) (2006). http://www.sbaoi.org/index1.htm. Accessed 27 Feb 2012

World Health Organization and Harvard Medical School (2009) Medicines use in primary in developing and transitional countries. http://www.who.int/medicines/publications/primary_care_8April09.pdf